D0962533

Water Resources Development and Management

Series Editors: Asit K. Biswas and Cecilia Tortajada

Asit K. Biswas
Cecilia Tortajada
Benedito Braga
Diego J. Rodriguez
Editors

Water Quality Management in the Americas

With 29 Figures and 49 Tables

AGÊNCIA NACIONAL DE ÁGUAS

Professor Asit K. Biswas
Third World Centre
for Water Management
Avenida Manantial Oriente 27
Los Clubes, Atizapan
Estado de México 52958
Mexico
Email : akbiswas@att.net.mx

Dr. Cecilia Tortajada
Third World Centre
for Water Management
Avenida Manantial Oriente 27
Los Clubes, Atizapan
Estado de México 52958
Mexico
Email : ctortajada@thirdworldcentre.org

Professor Benedito Braga
National Water Agency
SPS, Area 5
Quadra 3, Bloco M
70610-200 Brasilia, DF
Brazil
Email : benbraga@ana.gov.br

Diego J. Rodriguez
Environment Division
Sustainable Development Dept.
Inter-American Development Bank
1300 New York Avenue, N.W.
Washington, DC 20577
USA
Email : diegor@iadb.org

Cover photo: © Haroldo Palo, Jr.

Library of Congress Control Number: 2005932088

ISSN 1614-810X
ISBN 3-540-24290-2 Springer Berlin Heidelberg New York

Springer is a part of Springer Science+Business Media
Springeronline.com
© Springer-Verlag Berlin Heidelberg 2006
Printed in The Netherlands

The use of general descriptive names, registered names, trademarks, etc. in this publication does not imply, even in the absence of a specific statement, that such names are exempt from the relevant protective laws and regulations and therefore free for general use.

Typesetting: Camera ready by authors
Cover design: E. Kirchner, Heidelberg
Production: Almas Schimmel
Printing: Krips bv, Meppel
Binding: Stürtz AG, Würzburg

Printed on acid-free paper   30/3141/as 5 4 3 2 1 0

Preface

Considerable attention has been given during the past few years to the water crisis that many regions of the world may face in the coming decades. While the magnitude and the extent of the global water scarcity problems of the future should not be underestimated, a serious analysis of the current trends indicate that the main water crisis in the coming years is most likely to stem primarily from water quality deterioration and lack of investment funds, rather than from physical water scarcities per se, as is widely expected at present.

In spite of the seriousness of continuing water quality deterioration in most countries of the world, water quality management continues to be a somewhat neglected issue in the international water community. Even the magnitudes and extents of the water quality problems are not reliably known in many developed countries and in nearly all developing ones. Accordingly, in order to assess the current situation in water quality management in the Americas, and to discuss the alternatives available to improve the existing and future water quality conditions in a cost-effective and timely manner, the Third World Centre for Water Management in Mexico, the National Water Agency in Brazil (Agência Nacional de Águas), and the Inter-American Development Bank organised a workshop on "Water Quality Management for the Americas," in Fortaleza, Brazil, 12-13 April 2004. Participation to the workshop was by invitation only, and was restricted to 26 leading experts on the subject from the region. The participants came form different disciplines as well as institutions (federal and state governments, private sector, academia, NGOs, and intergovernmental organisations). The present book is based on the papers that were specially commissioned for this workshop and the resulting discussions.

Among the main objectives of the workshop were to examine the current status of water quality management in the Americas, assess the effectiveness of policies and programmes, analyse the impacts of using new and innovative techniques for managing water quality more efficiently, and to identify a set of best practices that can be used for the region. The workshop made a special effort to assess and synthesize knowledge and experiences in this area to determine which policies, management practices and techniques have worked and which ones have not, and the reasons why. Case studies were specifically commissioned from eminent experts from Argentina, Brazil, Costa Rica, Chile, Colombia, Mexico, Panama and the United States. In addition, the experiences of the North American Development Bank and International Food Policy Research Institute were also reviewed.

The issues that were discussed in-depth during this high-level workshop included, but were not restricted to, the current situations and future trends in water quality management in the world in general and in the Americas in particular; priority water quality issues for the region; appropriateness of legal, institutional, financial and policy frameworks; roles of river basin organisations as units for water quality management; and new economic instruments that could be used, including water rights and market-oriented approaches. Both point and nonpoint sources of pollution were considered.

The final report of the World Commission on Water had noted in 2000 that less than 10 percent of the contaminated water in Latin America was properly treated and disposed of in an environmentally-sound manner. The situation has not improved very much during the past five years. The health impacts and the social costs of these practices have yet to be properly assessed for any Latin American country, but their annual economic costs to the region are bound to be in billions of dollars. This means that water quality management in the region must receive significantly more attention than what it is getting at present, otherwise, improvements will continue to be incremental and marginal.

Comprehensive and long-term water quality data from much of the region are not available, and the reliability of a significant percentage of available data leaves much to be desired. Thus, unquestionably regular and reliable monitoring and evaluation of water quality conditions are prerequisites for efficient water quality management of the region. In addition, the data available should be easily accessible to anyone who requires that information, which at present is not the case for most of the region.

Because the institutional responsibilities for water quality management are fragmented at present, inter-institutional coordination and institutional strengthening and restructuring should receive special attention in the coming years. Furthermore, capacity building in technical and managerial aspects of water quality management is essential, as is reduction of political interferences and very significant improvements in the implementation of legal and regulatory regimes, and transparency of the administrative and management processes.

Some institutions have argued that water quality management would be more effective if it was implemented within the context of river basins, since it may be easier to obtain and analyse data, predict water quality trends of the future, and obtain reliable estimates of benefits to the society from increased investments in improving water quality within a specific geographical area. However, while this approach may be feasible, the main constraint at present is that the river basin organisations are not yet fully functional in most Latin American countries. Nor do experiences span over a significant period of time to draw definitive conclusions on their long-term effectiveness and impacts. Major efforts are thus needed to substantially improve the operational and management efficiencies of the existing river basin institutions, if water quality is to be properly managed within the framework of river basins. The Brazilian experience in the Paraiba River Basin has been encouraging thus far, but the length of this experience has been far too short to draw any long-term conclusion. Accordingly, for the Latin American region as a whole, the jury is still out as to whether river basins should be the most appropriate administrative unit for water quality management.

In terms of regional experiences on instruments for water quality management, their implementation depend primarily upon the appropriateness of the legal and institutional frameworks, and the existence of necessary technical and managerial capacities and political will. If the frameworks considered are not suitable for national and/or regional conditions, such instruments are likely to be of limited use. In fact, under certain conditions, they may even be counterproductive.

While economic instruments (bulk water charges, water rights, tradable permits, polluters-pay principle, incentives when necessary, etc.) can be of significant assistance to the region in improving the existing water quality management practices, a combination of economic instruments and command and control system is likely to provide an efficient approach to water quality management. Institutional innovations, as well as major strengthening of their capacities, are necessary. Legal and regulatory regimes need to be carefully tailored to meet the technical, social, economic and environmental requirements of the specific countries and regions. The laws and regulations are likely to be properly implemented on a regular basis only under such conditions.

In terms of financing, at present multilateral development banks are investing less than 5 percent of their funds for water and sanitation-related developments. Unfortunately, in most developing countries, Latin America included, conventional funds available for investments in the water sector (including water quality) are grossly inadequate. In other words, the public sector cannot provide all the necessary funds for the development and the management of the water sector, and the investments currently available from the private sector and the multilateral and bilateral funding agencies are not enough. Thus, the overall issue of how to meet the investment needs for the water sector (including water quality) needs urgent attention. Current trends indicate that the gap in funding, between what is needed and what is available, is likely to widen even further in the future. Thus, the funding requirements for water quality management, and how this funding can be obtained on a sustained basis in the future, are important issues that need priority discussion and resolution in the coming years.

In spite of this overall panorama, water quality management in most Latin American countries is not receiving the political attention and priority it deserves at national and regional levels. Simultaneously, overall governance, including political, legal and institutional conditions, has sometimes created an environment which is not favourable for new investments in the sector. In many countries, funding has significantly decreased at national levels, instead of increasing. For example, in Mexico, inspite of enormous efforts, the situation has become critical since the government at present can fund less than 30 percent of the needs of the country. Such investment shortfalls mean that the countries must become more and more efficient and innovative in the future in terms of generating new funds, and then use them more productively and efficiently than at present, or in the past.

Rational water resources management is simply not possible without proper water quality considerations. Furthermore, water quality problems are much broader than only construction and operation of wastewater treatment plants. Efficient water quality management on a long-term basis requires a much broader perspective. This could include, but not necessarily be limited to, important issues like formulation and implementation of national water policies, monitoring and evaluation of water quality conditions, presence of appropriate and functional legal and institutional frameworks, and capacity building programmes.

Control, regulation and impacts of nonpoint sources of pollution need considerably more attention that they are receiving at present. Even in an advanced country like the United States, after decades of sustained effort, nonpoint sources continue to remain a serious problem. For example, at present, agricultural activities are contributing to about 50 percent of pollution of the rivers in the United States. This indicates that nonpoint sources of pollution need more attention, if the current situation is to improve markedly in the coming years.

There is no question that point sources of pollution have received more attention in the past, primarily because, technically, economically, socially and institutionally, point sources are much easier to control than nonpoint sources. Fertiliser and pesticide use per unit of area is increasing steadily in most developing countries to increase agricultural yields. Thus, increasing attention needs to be paid to manage water pollution resulting from agricultural chemicals in the future.

In developing countries, a determined effort over the long-term is essential, if nonpoint sources of pollution are to be successfully controlled and managed. The effectiveness of point and nonpoint sources trading needs further debate. While in some countries it could be an important policy alternative, it may only be of limited use in some others, mainly because of their current socio-economic conditions. In other words, the payment for a service (e.g. environmental service) may not be feasible in the near future for some countries where people face serious problems to satisfy their own basic needs.

No one solution would address every concern of the region: there simply is no single magic bullet to solve all water quality problems of any specific country. Countries are heterogeneous in terms of physical, climatic, economic, social, institutional and environmental conditions. No one single paradigm is likely to be valid for all the countries of the American region. Use of approaches like command and control, public involvement, and use of economic instruments are all likely to be helpful under specific conditions. Equally, other considerations like formulation and implementation of country-specific water quantity and quality management policies and programmes may be necessary.

The workshop convened at Fortaleza was an unique event, where important major issues related to water quality management for the countries of the American region were objectively, comprehensively and critically examined, without any dogmas, consideration of vested interests, or political correctness. The effectiveness and impacts of different water quality management practices were objectively examined. The level and quality of the discussions during the workshop were consistently high. This is not surprising since a special effort was made to select and invite the best experts from the region.

The Third World Centre for Water Management is especially grateful to Prof. Benedito P.F. Braga, Director of Agência Nacional de Aguas of Brazil, who has assisted us consistently to formulate and implement many of our activities in Latin America. Without his technical and intellectual support, this project (and many others that the Centre has carried out in the past) could not have been implemented. The financial support of the Inter-American Development Bank is especially acknowledged, as is the support and encouragement we have consistently received from Diego Rodriguez and Silvia Ortiz of the Bank to complete this study. The help of Thania Gómez of our Centre to put the entire manuscript in the Springer format is much appreciated.

On behalf of the sponsors, we would like to express our gratitude and appreciation to all the authors of this book who accepted our invitation to prepare the definitive analyses of the specific issues and also to participate in the in-depth review and discussion during the Fortaleza workshop. Without their support, and commitments this book would not have been possible.

The present publication is a continuing effort by the Third World Centre for Water Management to synthesise the existing knowledge on water management from specific regions. The Centre is a knowledge-based organisation that specialises in generation, synthesis, application and dissemination of knowledge. For the Latin American region, the Centre has already completed authoritative analyses of water policies and institutions, integrated river basin management; women and water management; water pricing and public private partnerships in the water sector. Details of these and other books published by the Centre, and its current activities, can be seen in our website: www.thirdworldcentre.org.

Asit K. Biswas   Cecilia Tortajada
President   Vice President

Third World Centre for Water Management
Atizapán, Mexico

Contents

5. Integrated Water Quality Management in Brazil, by Benedito Braga, Monica Porto and Luciano Meneses

6. Institutional Aspects of Water Quality Management in Brazil, by Raymundo Garrido

7. Water Quality Management in Ceará, Brazil, by José Nilson B. Campos and Francisco de Assis de Souza-Filho

8. Water Quality Management in Mexico, by Felipe I. Arreguín-Cortés and Enrique Mejía-Maravilla

List of Contributors

Almeida-Jara, Raúl, Director General, Water Resources Development Department, Guanajuato State Water Commission, Autopista Guanajuato-Silao km 1, 36251, Guanajuato, Mexico

Arreguín-Cortés, Felipe I., Deputy Director General, National Water Commission, Insurgentes Sur No. 2416, Col. Copilco el Bajo, 04340, Mexico, DF, Mexico

Ballestero, Maureen, Coordinator, Global Water Partnership-Central America, Guanacaste, PO Box 14-5000 Liberia, Guanacaste, Costa Rica

Barrios, J. Eugenio, Consultant, Water Quality Management, Tezoquipa 44, 14090 Mexico, DF, Mexico

Blanco, Javier, Consultant, former Head of the Economic Analysis Office, Ministry of Environment, Housing and Territorial Development, Carrera 3, No. 74A-37 piso 2, Bogotá, Colombia

Braga, Benedito, Director, National Water Agency, SPS, Area 5, Quadra 3, Bloco M, 70610-200 Brasilia, DF, Brazil

Campos, José Nilson B., President, Brazilian Water Resources Association, Avenida Padre Antônio Tomás 3646 No. 1200, Fortaleza, Ceará 60190-080, Brazil

Clark, Christopher D., Assistant Professor, Department of Agricultural Economics, University of Tennessee, 321D Morgan Hall, Morgan Circle, Knoxville, TN 37996-4518, USA

Cline, Sarah A., Research Analyst, Environment and Production Technology Division, International Food Policy Research Institute, 2033 K. Street, N.W., Washington, D.C. 20006, USA

De Souza-Filho, Francisco de Assis, President, Ceará Foundation of Meteorology and Water Resources, Avenida Rui Barbosa 1246, Fortaleza, Ceará 60115-221, Brazil

Del Castillo, Lilian, Professor, Law School, University of Buenos Aires, Parera No. 36, Buenos Aires, C1014ABB, Argentina

Donoso, Guillermo, Associate Professor, Department of Agricultural Economics, Catholic University of Chile, Casilla 306, Correo 22, Santiago, Chile

Easter, K. William, Professor, Department of Applied Economics, University of Minnesota, 317g Classroom Office Building, 1994 Buford Avenue, St Paul, MN 55108, USA

Gallagher O'Neal, Suzanne, Advisor, North American Development Bank, 203 South St. Mary's, Suite 300, San Antonio, TX 78205, USA

García, Luis E., Consultant, 3031 Braxton Wood Ct., Fairfax, VA 22031-1337 USA

Garrido, Raymundo, Professor, Faculty of Economic Sciences, Federal University of Bahia, Praca 13 de Maio, 06-Piedade, 40070-010 Salvador, BA, Brazil

Guzmán, Zulma, Executive Director of the Latin American Association of Environmental Economists, Av. 82, No. 7-22 of 304, Bogotá, Colombia

Johansson, Robert C., Resource and Environmental Policy Branch, USDA-Economic Research Service, 1800 M Street, NW #S-4005, Washington, DC 20036, USA

Mejía-Maravilla, Enrique, Manager, National Water Commission, Insurgentes Sur No. 2416, Col. Copilco el Bajo 04340, Mexico, DF, Mexico

Melo, Oscar, Assistant Professor, Department of Agricultural Economics, Catholic University of Chile, Casilla 306, Correo 22, Santiago, Chile

Meneses, Luciano, National Water Agency, SPS – Area 5, Quadra 3, Bloco L, 70610-200 Brasilia, DF, Brazil

Porto, Monica, Associate Professor, Escola Politécnica, University of Sao Paulo, 05508-900, Sao Paulo, SP, Brazil

Reyes, Virgina, Technical Officer, Global Water Partnership-Central America, PO Box 14-5000 Liberia, Guanacaste, Costa Rica

Ringler, Claudia, Research Fellow, Environment and Production Technology Division, International Food Policy Research Institute, 2033 K. Street, N.W., Washington, D.C. 20006, USA

Rosegrant, Mark W., Director, Environment and Production Technology Division, International Food Policy Research Institute, 2033 K. Street, N.W., Washington, D.C. 20006, USA

Rodríguez, Raul, Managing Director, North American Development Bank, 203 South St. Mary's, Suite 300, San Antonio, TX 78205, USA

Russell, Clifford S., Professor Emeritus, Department of Economics, Vanderbilt University, and Research Associate, Bowdoin College, 15 Head Tide Church Rd. Alna, ME 04535, USA

Sandoval-Minero, Ricardo, Executive Secretary, Guanajuato State Water Commission, Autopista Guanajuato-Silao km 1, 36251, Guanajuato, Mexico

Water Quality Issues in Latin America

Luis E. García

1.1 Introduction

This chapter is an overview of some water quality issues that are relevant for Latin America. As a first step, some of the chemical properties so important in determining the quality of water are reviewed in order to understand the concept and meaning of the term "water quality."

A natural system's quality interchanges are then reviewed and compared to interchanges that take place in an antrophic system, and the relative nature of the water quality concept, vis-à-vis the beneficial uses of the water resource, is stressed.

The functional relationship between the concentration of substances in water and the streamflow is illustrated by a simple mass balance example. This is also used to stress the advantages (and difficulties) of applying the concept of receiving waters criteria vs. end-of-pipe standards for pollution control.

Some of the world's major water quality issues are reviewed as a preamble to the discussion of Latin American issues. The conclusion is that these issues relate to a lack of wastewater treatment, financial constraints (due to the high cost of treatment), difficulties in applying receiving waters criteria vs end-of-pipe standards, and finally a shortage of water quality data and monitoring programmes. Efficient basin-wide approaches to circumvent the high cost of treatment vs. the traditional project-by-project approach, although proposed by some multilateral financial institutions, are not yet applied on an operational basis.

Before moving on to the main theme of this analysis, i.e. the major quality issues in Latin America, it would perhaps be convenient to refer to some basic concepts about the raw material we are dealing with: water. These concepts are well known and can be found in many publications and pages on the Internet (e.g. The Water Page). But they are sometimes forgotten or dismissed in today's highly technical and conceptual world of economic, environmental and socio-political issues.

The chapter will touch on only some of the main characteristics that make water a unique inorganic composite and which are considered relevant to the topic of water quality. Therefore, the relevance of water as a precious resource for all living species on Earth and its importance for all human economic and social activities will not be addressed. Neither will the importance of the hydrologic cycle, which makes water the most vital of all substances upon which all life on Earth depends. The main subject to be addressed is its chemical composition and its effect on what people refer to as "water quality." First, as everybody learns at elementary school, two positive hydrogen atoms linked to one negative oxygen atom form the water molecule. This chemical bond is angular in shape, with form the water molecule. This chemical bond is angular in shape, with negative and positive charges on opposite sides.

Thus, the water (H_2O) dipolar molecule has a strong tetrahedral arrangement that needs much more energy to break it than other similar molecules with two hydrogen atoms of progressively higher molecular weight, such as hydrogen sulphide (H_2S), hydrogen selenide (H_2S_e) and hydrogen telluride (H_2T_e). Water, whose molecular weight is 18, boils around 100^0C, while the boiling point of the other three, with molecular weights of 34, 81 and 130, respectively, is found at – 61°C, -42°C and –4°C, respectively.

Moreover, water's freezing point is 0°C and for the other three it is –82°C, -64°C, and –51°C, respectively. This makes possible the presence of water in all three (liquid, gaseous and solid) states at the prevailing temperatures on the surface of the Earth and consequently the hydrologic cycle and, as a result of this, the existence of plant and animal life on the planet.

As regards water quality, it is important to remember that the strong molecular bond of water and its dipolar charge "break" other weaker molecules, such as NaCl, whose positively charged Na atoms are attracted by the negative charge of the water molecules.

Similarly, the negatively charged Cl atoms will be attracted to the positive charge of the water molecules, causing ionization. This makes water a potent natural "solvent", breaking up molecules of the substances it comes in contact with and incorporating their ionized atoms. Water in movement, either in the atmosphere or on the Earth's surface, also carries away gases and suspended matter, both conditions giving origin to what it is known as "water quality."

1.2 Water Quality Revisited

1.2.1 Natural Water Quality Interchanges

It is not hard, therefore, to reconcile the fact that in nature there is no such thing as "pure" water. Such a condition is only attained in laboratory-distilled water.

A continuous quality interchange takes place in nature as the hydrologic cycle proceeds. Water changes state and moves from the atmosphere to the Earth's surface, which it runs across, infiltrates through to lower strata, moves to the ocean and other freshwater bodies and returns to the atmosphere through evaporation and evapotranspiration.

As water arrives from the atmosphere through precipitation (P), it carries with it dissolved gases, dust and smoke particles, bacteria, salt nuclides and dissolved solids to the land's surface and to fresh and saline water bodies. When it comes into contact with land, it collects biochemical, unstable matter from live animals and from dead plants and animals. All this organic debris, as well as soluble and particulate products of biodegradation, plus silt, silica, mineral residues, bacteria, dissolved gases, soil materials and dissolved minerals from surface debris and primary rocks, is transported to fresh water bodies, both surface and underground, by runoff (R) and infiltration (I).

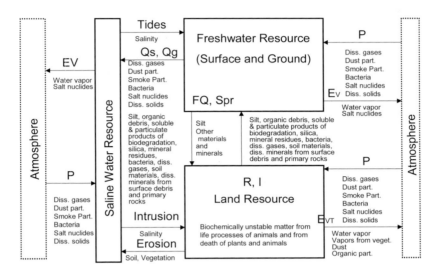

Fig. 1.1. Quality Interchanges in a Natural Hydrological Cycle (Source: Author notes from the Department of Civil Engineering, Hydraulic and Sanitary Engineering Division, University of California, Berkeley)

The accumulated sum of all of the above is carried and discharged by flowing surface (Q_s) and groundwater (Q_g) into the ocean. Soil and vegetation is also directly transported to fresh and saline bodies of water through erosion of land surfaces.

Silt and other materials and minerals carried by surface and groundwater are deposited back on the land surface by springs (Spr) and floods (FQ) and salinity is increased in land and fresh water resources by ocean intrusion and tidal effects.

Water evaporates directly from inland water bodies and from the ocean (EV), carrying water vapour and salt nuclides back to the atmosphere and vapours from vegetation, silt, and organic particles through evapotranspiration (EVT) from vegetation, completing the natural cycle. This cycle, of course, varies annually and seasonally according to the prevailing natural precipitation and hydrologic regime and according to the nature and state of the soil cover, natural vegetation and type of underlying basic rocks and aquifer material of the watershed.

1.2.2 Water Quality Interchanges Modified by Human Use

Land usage in the watershed and water use for human purposes introduce new water quality interchanges to the natural cycle described above. When precipitation reaches a given point on the land surface, a water balance is established. This involves the water absorbed by plants, the water that flows through the topsoil, that which is taken by plant roots and returned later by these to the atmosphere through evapotranspiration, that which infiltrates to lower strata to form aquifers and that which flows overland through gravity.

Any kind of human activity, such as urbanisation, tree cutting or planting, plant cultivation, water extraction for hydropower or human consumption, wastewater disposal, road building, etc., will alter these balances at any point in the watershed. Both the liquid and solid flow of a stream at a given point, the substances carried by water and the flow velocity reflect the regime of upstream precipitation, the characteristics of the river basin and of the human use of land, water and other natural resources. Any change in the contributing watershed is reflected in streams. At the same time, the reaction of these so as to adjust to the new situation will affect the downstream basin (García and Quiroga 2000).

Land use, water extraction and flow regulation modify the natural water balance in the watershed. The quality features of flows returning from beneficial uses are very different from those of extracted flows. Any change, however, will not be "good" or "bad." It will simply have consequences of which we should be aware. Figure 1.2 illustrates these changes simply from the water quality standpoint, considering only domestic, agricultural, and industrial uses. The second example, illustrated by this figure, although simplified, may be closer to reality, as human activity has indeed become part of the hydrologic cycle.

Water is extracted from surface and ground freshwater sources as well as from saline water sources (R) for all beneficial uses. These sources will have their natural quality modified by human activity. Additional modifications will be introduced to the extracted volumes by water treatment in order to make them suitable for the intended use, such as domestic or industrial. The quality will be the same, minus the treatment, for both unused surface and ground waters flowing directly into the ocean or other bodies of water.

Some water will be consumed during use and some will be returned to the bodies of water. Return flows from beneficial uses (RF) to surface and ground water sources and to the ocean will have different quality features, depending on their usage. Return flows from agricultural use will carry principally salts, nutrients, pesticides and organic debris. The main characteristics of industrial return flows will be organic matter, metal ions, chemical residues, higher temperature and increased salt concentrations. This of course is an oversimplification, as they will vary according to the type of industry and manufacturing processes. Domestic wastewater will carry mainly waste left by humans, such as garbage, grease, detergents, dissolved solids, bacteria, viruses and some industrial waste that may be deposited in sewers. All water consumed domestically, by plants and by industrial processes (C) will have the increased salt concentrations of return flows.

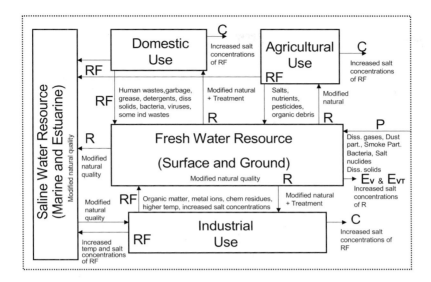

Fig. 1.2. Quality interchanges in Modified Hydrological Cycle (Source: Author notes from the Department of Civil Engineering, Hydraulic and Sanitary Engineering Division, University of California, Berkeley)

1.2.3 Some Working Concepts

From the above discussion, it follows there is no "good" or "bad" water quality. Water quality is a relative term that should be associated with a given water usage and its benefits, including environmental and ecosystem benefits. It has to do with the physical, chemical and biological features of a given body of water in relation to its suitability for a particular usage. Thus, a body of water might be of "good" enough quality for agriculture, but of "bad" or inadequate quality for a given industrial process. Another could be "good" for industry but "bad" for human consumption, and so on.

In the United States, it is usually said that a given body of water is polluted when it has substances that make it inadequate for a given beneficial use, and that it is contaminated when it has substances that endanger human health. Thus, a particular body of water may be polluted but not necessarily contaminated. However, if it is contaminated, it is polluted. In Latin America that distinction is not usually made and the terms "polluted" and "contaminated" are sometimes used muddled up. Although "contaminated" is the correct term since in Spanish, the word "pollution" has a completely different meaning.

All substances are potential pollutants. It depends on their concentration. All pollutants have the potential to become contaminants if their concentration is high enough to imperil human health. It must not be overlooked, however, that the concentration is a function of the streamflow of both the body of water's discharge and reception, so it varies with the hydrologic regime.

This is important for drafting discharge regulations or legislation. A regulation or legislation that prohibits water disposal of "any polluting substance" or even "any contaminant" would be meaningless if it were not accompanied by the corresponding limits on concentration. This would also be true for applying limits to substance volumes instead of concentrations, except in the case of closed bodies of water that may be looked upon as "sinks."

1.3 Standards vs. Criteria

When regulatory bodies refer to concentrations, two eventualities may arise: the concentration limits may refer to the discharge and they may also refer to the receiving body. The former are usually referred to as "end-of pipe standards," while the latter are usually called "receiving waters criteria." It is not difficult to imagine that the application of the former is tailored to point out source pollution, such as domestic or industrial wastewater discharges, while the latter are broader in scope. Figure 1.3 illustrates a simple case to show the rationality and advantages of "receiving waters criteria" over end-of-pipe criteria:

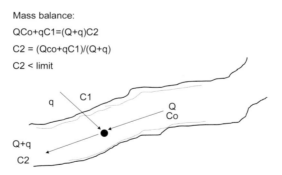

Fig. 1.3. Advantages of receiving water criteria

A point source has an effluent discharge q with a certain substance concentration C_1. The receiving stream has a flow Q and a concentration Co of the same substance. Assume there is a complete and immediate mix at the point of discharge: the downstream flow will be (Q+q) and the substance concentration will be C_2. The end-of-pipe standard approach will place a limit on the discharge con-

centration q, disregarding the diluting capacity of the stream, requiring more effluent treatment and probably incurring higher costs. The receiving waters criteria
approach will place a limit on the resulting substance concentration C2 of the receiving waters, taking full advantage of the diluting capacity of the stream, requiring less treatment of the effluent and, possibly, cost less.

Although not always the case, when implementing effluent discharge regulations most countries in Latin America opt for the end-of-pipe standards approach.
It is believed the reason is twofold: first, a successful implementation of the receiving waters criteria requires effective monitoring and a substantial amount of
data. Second, often the major concern is industrial pollution, and the end-of pipe
standards approach allows for more direct, less expensive and easier operation implementation procedures.

However, some countries like Mexico are implementing the receiving waters
approach and some, like Brazil and Chile, are implementing a mixed approach.
Mexico developed a Water Quality Index (ICA by its Spanish acronym) to classify
bodies of water according to their degree of pollution (CNA/SEMARNAT 2004).
The ICA is a weighted average of 18 quality indexes, one for each of the same
number of water quality parameters considered to be of importance. Figure 1.4 illustrates the ICA calculation.

$$ICA = \frac{\sum\limits_{i=1}^{n} I_i W_i}{\sum\limits_{i=1}^{n} W_i}$$

$i = 1, 2, ..., 18$ each one of 18 water quality parameters considered

$n = 18$

I_i = quality index for parameter i

W_i = Relative weight given to parameter i

SOURCE: Author notes from the Water Quality Division, National
Water Commission of Mexico

Fig. 1.4. Water Quality Index. ICA ranking: 0-29% highly polluted; 30-49% polluted; 50-
69% slightly polluted; 70-84% acceptable; 85-100% non polluted

The ICA ranking was used to classify bodies of water and there was, of course,
a different classification for each one of the five uses that were considered: public

water supply, recreation, fishing/aquatic life, industry and agriculture. As of 2004, however, the use of the ICA was suspended while new criteria for water quality evaluation were developed, aligning them to a greater degree with the conditions of the national water quality-measurement network (CNA/SEMARNAT 2004). Meanwhile, only two parameters, related to the pollution caused by the country's major urban and industrial centres, are being used: 5-day biochemical oxygen demand (BOD_5) and chemical oxygen demand (COD).

Every country that establishes water quality criteria does so according to the water usage of importance to the country or specific regions within the country. For example, more than 40 years ago (McKee and Wolf 1963) the state of California adopted raw water source criteria for nine beneficial uses. Some of these criteria have also been utilised in other parts of the US. and in other countries and have undergone many revisions and modifications over the years. They are mentioned here, however, only as an example of the relativity of the "water quality" concept and its intimate relationship to beneficial uses, and as a way of demonstrating that this concept is not, by all means, new:

- Domestic water supply: these included BOD_5, coliform MPN per 100 ml, dissolved oxygen, pH, chlorides, fluorides, phenolic compounds, colour, and turbidity.
- Industrial water supply for more than 40 industries.
- Irrigation, including sodium (% sodium and sodium adsorption ratio SAR), specific conductivity (concentration of ions), total salts, boron, chlorides, and sulphates.
- Stock and wildlife watering: it is assumed that water that is safe for humans is also safe for stock, with added threshold salinity concentrations and control of nitrates, fluorides and selenium and molybdenum salts.
- Propagation of fish and other aquatic and marine life: these include dissolved oxygen, pH, specific conductance, free carbon dioxide, ammonia and suspended solids.
- Shellfish culture, including temperature, salinity, pH, and dissolved oxygen.
- Swimming and bathing waters: these include coliform MPN per 100 ml, number of illnesses per 1,000 persons-days, bacteria per 100 ml, that they be aesthetically enjoyable and must contain no substances that are toxic upon ingestion or irritating to the skin.
- Boating and aesthetic enjoyment: waters must be aesthetically enjoyable and be free of excessive inorganic nutrients (such as nitrates, phosphates, carbonates or silicates) that hasten the eutrophication of lakes and streams.
- Hydropower and navigation: these include substances (such as acid, alkali and excessive salinity) that accelerate corrosion and cavitation, materials (such as debris, silt and other suspended solids) that block channels and intakes, organic matter that generates putrescible odours and corrosive hydrogen sulphide gas in deep reservoirs, algae, fungi, worms, barnacles and other pollutants that clog passageways and cling to vessels or accelerate corrosion, borers that destroy wharves and docks, and floating oil that is a potential fire hazard.

1.4 Global Major Water Quality Issues

To highlight the importance of water pollution all over the world, the United Nations has recently compared what is considered the "average natural concentration" of nine substances in river waters with what is considered the "actual concentration" of the same substances (United Nations 2003).

These substances are: calcium (Ca^{++}), magnesium (Mg^{++}), sodium (Na^+), potassium (K^+), chloride (Cl), sulphate (SO_4), bicarbonate (HCO_3), silicon dioxide (SiO_2), and totally dissolved solids TDS. Why these substances were selected or how the average natural concentrations were determined, given that "over the last 200 years human activities have developed to such an extent that there are now few examples of natural water bodies," is beside the point.

What is important is that increasing attention and concern is being given across the world to water quality issues beyond those related merely to water quantity. Global problems like heavy metals have been identified. There are also regional problems, caused by acid rain and local groundwater problems.

Furthermore, the U.N. identifies the world's major water quality issues as being the following (United Nations 2003): organic pollution, pathogens, salinity, nitrate, heavy metals, acidification, eutrophication and sediment load (increase or decrease); organic pollution, pathogens, and nitrate in lakes; salinity, nitrate, eutrophication and sediment loads in rivers; and organic pollution, pathogens, and acidification in groundwater.

These are all deemed a serious issue on a global scale. Organic pollution, pathogens and acidification in rivers, salinity and nitrate in groundwater and acidification and eutrophication in lakes are considered a very serious issue on a global scale.

Organic matter from domestic sewage, municipal waste and agro-industrial effluents that are discharged untreated is the most widespread pollutant globally, particularly in Asia, Africa and Latin America. Increased salinity is also a serious form of water pollution in the developing world. For the developed world, acidification of surface waters was a problem some three-four decades ago but has since diminished with the control of sulphur emissions in Europe and North America.

1.5 Major Water Quality Issues in Latin America

Several authors concur with the U.N. evaluation of the major sources of pollution in Latin America (United Nations 2003). Drawing from various sources, Russell et al. (2001) cite forecasts of increases of about 125 percent in annual wastewater volumes from 60.3 km^3/year in 1990 to 134.3 km^3/year in 2025. Most of these increases were expected to be in industrial and municipal wastes.

No significant increase in agricultural return flows was forecast at the time. According to FAO sources (FAO 2004), significant water pollution due to irrigation has nevertheless been reported in Barbados, Mexico, Nicaragua, Panama, Peru, Dominican Republic and Venezuela. The FAO also points out that secondary salinity, i.e. induced by irrigation, is a serious constraint in Argentina, Cuba, Mexico, and Peru and, to a lesser extent, in the arid regions of north-eastern Brazil, north and central Chile and some small areas of Central America. Other sources such as the OECD (2003) refer to water quality impacts induced by subsidies in agriculture, as summarised in Table 1.1.

Table 1.1. Water quality impacts caused by subsidies in agriculture

Subsidy	Agent of impact	Impact
Agricultural price support policies	Incentive for water-inefficient crops	Salinisation, waterlogging, decline in groundwater
Surface water price	Overuse of water	Pollution, salinisation
Electricity price	Substitution of surface water for groundwater	Aquifer depletion, salinisation
Pesticide price	Overuse and inefficient use of pesticides	Surface water and groundwater contamination
Fertiliser price	Overuse and inefficient use of fertiliser	Surface and groundwater contamination

Source: OECD, 2003.

This is a problem affecting developing and developed countries alike and the OECD points out that globally farmers rarely pay more than 20 percent of the real cost of water.

Information on water-borne diseases linked to irrigation is scarce and difficult to obtain, and it is even more difficult to distinguish whether these diseases are caused by irrigation or by the presence of wet zones, marshes, areas susceptible to waterlogging or simply pollution (FAO 2004). However, their incidence among the population of some Latin American countries cannot be overlooked. Water quality problems, caused by aquifer marine intrusion, has also been reported as significant in parts of Mexico, Venezuela, Dominican Republic and the Antilles.

Russell et al. (2001) quote sources that estimate that 30 million m³ of domestic wastewater are discharged everyday into Latin American bodies of water. This adds a load of over two million tons of BOD per year to the receiving waters. Approximately 70% of Latin America's 337 million urban inhabitants have adequate sanitation, but more than 100 million urban dwellers do not. These are concentrated in countries with large cities, such as Brazil (57 million have inadequate sanitation), Mexico (10 million), Argentina (8 million), Venezuela (7 million) and Peru (7 million).

Large percentages of sewage are disposed untreated into the bodies of water. One source (IADB 2003a) estimates that the overall percentage of persons covered with wastewater treatment in 2000 was as follows:

Over 50%: Barbados (100%), Antigua and Barbuda (95%), Bahamas (80%), Uruguay (73%) and Trinidad and Tobago (64%).

From 50% to 10%: Guyana (49%), St. Lucia (44%), Belize (40%), Dominican Republic (34%), Nicaragua (32%), Bolivia (26%), Panama and Cuba (18%), Chile (16%), Mexico (14%), Peru (11%), Colombia (10%) and Surinam (10%).

Countries with less than 10%: Argentina (9%), Brazil and Paraguay (8%),

Venezuela (7%), Ecuador (5%), Costa Rica (4%), Honduras (3%), El Salvador (2%) and Guatemala (1%).

Countries with less than 1%: Haiti, Dominica, Granada, Saint Kitts and Nevis and St. Vincent and Grenadines Islands. Some of these figures may be subject to dispute or rectification by individual countries. However, altogether they serve to emphasise the magnitude of the problem.

There are many reasons for the above percentages. Russell et al. (2001) cite four: First, the high cost and complexity of treatment facilities is beyond the means of most communities. Second, foreign aid agencies and local governments have given priority to water supply projects rather than sewage treatment. Third, it is easier to charge for water than for wastewater treatment; and fourth, partial solutions that removed human waste from towns and cities left the treatment part of the equation for later, and it was often not completed.

This situation has created major downstream pollution problems at municipal and domestic levels in several countries' rivers, such as one in Mexico's Mezquital Valley that receives some 50 m³/sec of raw sewage from Mexico City, the Reconquista in Argentina, Brazil's Tietê and Paraiba Rivers, Colombia's Bogotá River, Costa Rica's Grande de Tárcoles, Guatemala's Las Vacas-Motagua Rivers and El Salvador's River Sucio. All of these have a high level of dissolved oxygen (DO) depletion and bacterial or toxic substance contamination.

Russell et al. (2001) also quote several authors as saying that western environmentalists have been preoccupied with the preservation of tropical forests and endangered species and have ignored the problem altogether. These authors are also quoted as blaming the problem on the high birth rates in the countries of the region and as claiming that local populations have shown a willingness to pay for wastewater collection and disposal but not for treatment.

However, in a recent survey (IADB 2003b), conducted among 1,500 sanitation services officials and investors, the highest investment priority in the water and sanitation sub-sector for the 400 people who completed the questionnaire they were given was sanitation – including wastewater treatment. This demonstrates that, among municipal and domestic sources, pollution is the region's major overall water quality issue. The results of the survey are summarised in Table 1.2.

Table 1.2. Priority for investments

Activity Priority	Very high	High	Medium	Low	Unnecessary
Water production and distribution	35	34	24	6	1
Sewerage and sewage treatment	59	28	11	2	0
Unaccounted for water and loss reduction	38	35	22	5	0
Metering	27	40	21	9	3
Distribution to low-income neighbourhoods	39	33	23	5	0

Source: IADB, 2003.

The significance of the above is even more relevant, given that the Millennium Development Goals refer to access to water and sanitation services but do not specifically refer to wastewater treatment. If the goal of reducing by 50% the population without access to wastewater treatment (IADB 2003a) by 2015 is to be reached, Latin America will have to make a concentrated effort, as shown in Figure 1.5.

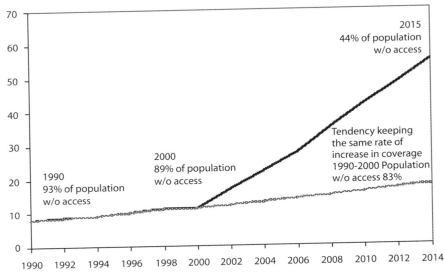

Fig. 1.5. Development goals: percentage of population with access to wastewater treatment

To attain this goal, a total investment of $17.7 billion will be required between 2000 and 2015, or $1.2 billion annually (IADB 2003a). Country by country, this would vary from a high of $479 million per year in Brazil, $181 million per year in Mexico and $136 million per year in Argentina to a low of $1-2 million per year in the Antilles and $5 million per year in Uruguay and Panama, with all the other countries falling somewhere in-between.

For the larger economies, this may not represent too big an effort but for smaller economies such as the Dominican Republic, Haiti, Guatemala, Paraguay and Ecuador investing $14-28 million per year for wastewater treatment may indeed be colossal. It may well be that the main constraint, among those mentioned by Russell et al. (2001) may be the one related to the high cost of treatment.

1.6 Basin-Wide Approach to Water Quality Management

IADB's strategy for Integrated Water Resources Management (IADB 1998) calls for water quality investments to be made following a watershed or river basin approach. It also calls for these investments to be considered not in isolation but integrated with other water resources management investments.

It favours the watershed or river basin approach, among other reasons, because it allows for the conceptualisation of comprehensive programmes that allow for a more efficient use of financial resources, a reasonable compromise between the different beneficial uses of water and a better management of both the volume and quality of the available water resource. It does not recommend how the economic analysis of investments to improve ambient water quality should be done, nor does it explicitly mention or require the optimisation of net benefits.

The large investments needed for complying with the corresponding Millennium Development Goals of reducing the number of persons without access to wastewater treatment by half by 2015 represent an important constraint imposed on many Latin American countries.

In the face of water quality problems caused by municipal and domestic pollution, Russell et al. (2001) explored the possibility of applying not traditional project-by-project investment but the watershed or river basin approach to solve the problem. They concluded, however, that there were three major hurdles: It would require: 1) The collection of a large amount of basin-wide reliable data 2) The availability of predictive water quality models and 3) An estimate of the investments needed to obtain benefits from water quality improvement basin-wide.

The first hurdle, although formidable, could be overcome and the second was not a major problem, since such models existed and had been applied. The major difficulty was the third hurdle, as estimating benefits is very difficult. Thus, the critical limitation on achieving a basin-wide, optimum water quality is mainly related to economics (Russell et al. 2001). Given these limitations, Russell et al (2001) proposed a procedure, which they called "second-best." Even if it does not guarantee an optimum balance will be obtained, at least it will satisfy the spirit of IADB's strategy for IWRM. This procedure takes an integrated basin-wide approach in a lowest-cost sense and does cost-benefit analysis as a second step to compare the costs of attaining water quality goals with the benefits of those goals. Assuming that the political (or government participation) process has chosen the desired receiving waters quality criteria, or range of standards, the locations, designs and capacities can be identified in a given basin to minimise the cost of such in-stream water quality criteria at the lowest possible cost (Russell et al. 2001).

However, the IADB is not yet applying this approach and the project-by-project way of doing things to meet end-of-pipe standards continues.

1.7 Conclusions

Despite worthwhile efforts across Latin America, many water quality issues prevail in the region. Some of these issues are conceptual in nature, others are financial and others are of a practical nature.

Not all the countries in the region face the same water quality problems. They vary according to the stage of development and the type of economic activity that dominates in a given geographical area or country. It can be said, however, that a major water quality problem common to all countries is the disposal of raw sewage in receiving bodies of water and the lack of wastewater treatment. Associated with this is the high cost of treatment, making it very difficult for many countries to comply with wastewater treatment goals equivalent to the Millennium Development Goals.

More often than not, attempts are made to circumvent these problems by establishing end-of-pipe standards, but the end result has not always been desirable. Moreover, this approach has been geared mostly towards industrial pollution, ignoring municipal and domestic pollution. A general lack of water quality data and monitoring efforts has also precluded the development and successful application of receiving waters criteria, except in a few countries.

This lack of in-stream quality criteria also precludes the application of more efficient basin-wide approaches that try to cope with the financial limitations for adequate project-by-project approaches.

1.8 References

CNA/SEMARNAT (2004) Estadísticas del Agua en México. Comisión Nacional del Agua/Secretaría de Medio Ambiente y Recursos Naturales, México

FAO (2004) AQUASTAT Information System on Water and Agriculture http://www.fao.org/ag/agl/aglw/aquastat/regions/lac/print6.stm

García L, Quiroga R (2000) Consideraciones Institucionales para Abordar Bienes Públicos Regionales: Programa Trinacional de Desarrollo Sostenible de la Cuenca Alta del Río Lempa. Regional Public Goods & Regional Development Assistance Conference. IADB, ADB, USAID, Washington, DC

IADB (2003a) Workshop on Financing Water and Sanitation Services: Options and Constraints. Inter-American Development Bank, Washington, DC

IADB (2003b) Workshop on Financing Water and Sanitation Services: Options and Constraints. Inter-American Development Bank, Washington, DC

IADB (1998) Strategy for Integrated Water Resources Management. IADB Strategy Paper No. ENV-125. Inter-American Development Bank, Washington, DC

McKee JE, Wolf HW (1963) Water Quality Criteria, 2nd edn, Publication 3-A. State Water Quality Control Board. The Resources Agency of California, Pasadena, CA

OECD (2003) Working Party on Global and Structural Policies. Working Group on Economic Aspects of Biodiversity. Perverse Incentives in Biodiversity Loss. ENV/EPOC/GSP/BIO(2003)2/FINAL, Paris

Russell CS, Vaughn WJ, Clark CD, Rodriguez DJ, Darling AH (2001) Investing in Water Quality, Measuring Benefits, Costs and Risks. Inter-American Development Bank, Washington, DC

The Water Page. http://www.thewaterpage.com/waterbasics.htm

United Nations (2003) Water for People, Water for Life. The UN World Water Development Report, Barcelona

Economic Instruments and Nonpoint Source Water Pollution

Clifford S. Russell and Christopher D. Clark

2.1 Introduction

Nonpoint source (NPS) water pollution is generally agreed to be the dominant cause of ambient water quality problems in the developed (OECD) world. It has achieved this distinction primarily by escaping the increasingly strict and sophisticated regulation that has reduced the flow of point source (PS) pollution over the last 30 years[1]. Over the same period, increases in farming intensity have also led to increases in both the application of chemical fertilisers and pesticides and the production and agglomeration of livestock wastes[2]. Both have resulted in increased runoff and leaching of pesticides and nutrients into neighbouring water bodies[3]. Table 2.1 illustrates the relative importance of agricultural pollution sources in general, and of the specific pollutants, nutrients and silt, in the impairment of United States waters. Figure 2.1 traces the growth of fertiliser application in the United States from 1961 to 2001. Figures 2.2 and 2.3 illustrate the growth in "excess nutrients" (i.e., quantities in excess of the potential uptake of the farm's crops and pastures) generated by the manure produced by confined animal opera-

---

[1] Overviews of the effect of PS control in the US can be found in Adler et al. (1993) and USEPA (1989).

[2] Managed forests (plantations), golf courses, road and other construction sites, and even city streets and suburban parking lots may be lumped under the NPS label and may be important causes of local water quality problems. We shall concentrate on agriculture, however, as being the largest problem seen at the national level, and likely the most relevant to developing nations, now and in the intermediate run future.

[3] "All else being equal, the highest efficiency of nitrogen fertiliser is achieved with the first increments of added nitrogen; efficiency declines at higher levels of addition. Today, only 30–50% of applied nitrogen fertiliser and ~45% of phosphorus fertiliser is taken up by crops. A significant amount of the applied nitrogen and a smaller portion of the applied phosphorus are lost from agricultural fields. This nitrogen contributes to riverine input into the North Atlantic that is 2- to 20-fold larger than in preindustrial times. Such nonpoint nutrient losses harm off-site ecosystems, water quality and aquatic ecosystems, and contribute to changes in atmospheric composition. Nitrogen loading to estuaries and coastal waters and phosphorus loading to lakes, rivers and streams are responsible for over-enrichment, eutrophication and low-oxygen conditions that endanger fisheries." (Tilman et al. 2002: 673).

tions in the United States from 1982 to 1997[4]. Given the magnitude of NPS pollution's contribution to water quality impairments[5], it is not surprising that there is widespread agreement that many ambient water quality goals cannot be reached without reducing NPS pollution (Ribaudo et al. 1999). Somewhat more surprising is the extent of the agreement on the cost-effectiveness of NPS pollution control relative to further tightening of PS regulations, such as requiring tertiary treatment of municipal and industrial wastewater, aimed at removing nitrogen (N) and phosphorus (P) (Faeth 2000).

This last observation raises in fairly stark terms the question of why OECD countries have not regulated NPS sources with more vigour. One reason seems obviously to be conflict with national agricultural policies and the political power of the agriculture sector, which is out of scale with its importance in OECD economies, as measured, for example, by sectoral employment or value added. But that is not the entire story. Another important reason, and one central to the message of this chapter, is that NPS water pollution poses uniquely difficult problems for would-be regulators. Most importantly, the technology that would allow actual measurement of discharges from any single economic unit (farm) does not now exist. This makes the familiar arsenal of policy instruments for pollution control of very dubious value. Thus, writing permits that purport to set upper limits on pollution discharges per unit time would be pointless – mere statements of aspiration with no practical effect – because compliance by a permitted source could not be established. By the same argument, the favourite instruments of economists -- effluent charges and tradable discharge permits – are of limited value, at least as they are commonly described and discussed. Finally, the practical burden of regulating NPS discharges by any source-specific method would be very large. For example, in the United States there are currently more than 5 times as many farms as there are permitted PS dischargers[6].

---

[4] Farm-level nutrient balances are important because, to minimize the costs of their disposal, "excess nutrients" are often applied at rates above those the crops can absorb.

[5] For example, a study of 86 US rivers found that NPS accounted for over 90% of the nitrogen inputs in over one-half of the rivers studied and for over 90% of the phosphorus inputs in one-third of the rivers (Newman 1995).

[6] However, the gap, in the US anyway, is closing. At the time the CWA was enacted, there were approximately 2.86 million farms and only 100,000 PS subject to the permitting requirements, or over 28 farms for every point source (USEPA 2001). Since then, farm numbers have plunged to about 2.16 million, while the number of permitted PS has increased to 400,000 and is projected to increase to 500,000 or more as a result of changes in regulations to incorporate large construction sites, storm water systems, and confined animal feeding operations within the permitting programme (USEPA 2001; USDA 2003).

Table 2.1. Percentage of water quality impairments contributed to by selected pollutants and sources of pollutants

	Percentage		
	River	Lake	Estuary
Pollutants			
Metals	15	42	52
Oxygen-depleting substances	21	15	34
Nutrients	20	50	
Bacteria	35		30
Siltation	31	21	
Pesticides		8	38
Total dissolved solids		19	16
Sources of pollutants			
Agriculture	48	41	18
Urban runoff/storm sewers	13	18	32
Municipal point sources	10	12	37
Hydrologic modification	20	18	14
Atmospheric deposition		13	24
Resource extraction	10		12

Source: USEPA, 2002, updated and adapted from Faeth, 2000.

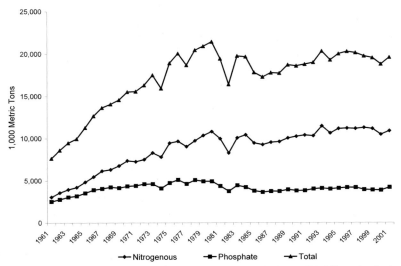

Fig. 2.1. Fertiliser consumption trends in the United States (Source: FAOStat, Agricultural Data)

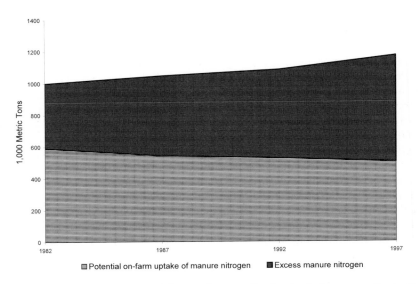

Fig. 2.2. Total United States manure nitrogen from confined animals in excess of potential on-farm uptake (Source: Adapted from Gollehon et al., 2001)

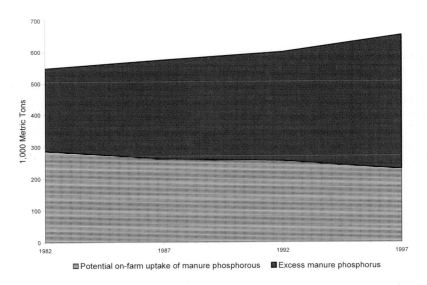

Fig. 2.3. Total United States phosphorous from confined animals in excess of potential on-farm uptake (Source: Adapted from Gollehon et al., 2001)

It is also worth noting that the three considerations just sketched are interrelated in a way that we believe makes the influence of the complex greater than the sum of its parts. Thus, the inability to monitor, implying the loss of common tools of pollution management, leads more or less ineluctably to suggestions for tackling the problem that are arguably more intrusive and, thus, more offensive to farmers than would be those tools. Consider: a discharge permit, even one drawn up, as under the United States Clean Water Act (CWA), on the basis of an assumed discharge reduction technology, does not require employment of that particular technology. And the common choices for meeting the permit terms tend to be end-of-pipe treatment units that impose few requirements on upstream process actions (for industry) or household decisions (for municipal sources). But to manage agricultural pollution, the recommendations of voluntary programmes in OECD countries typically consist of so-called best management practices (BMPs) that explicitly specify changes in the way agricultural production is undertaken. Examples include switching from conventional tillage to no-till cropping, and transforming parts of fields bordering watercourses into buffer strips or wetlands[7]. It does seem likely that such changes, viewed as potential intrusions, would be strongly resisted by the agricultural community. Special environmentally-justified fertiliser taxes are another common suggestion[8]. These might reasonably be seen as less intrusive, if still painful, especially if uniform across a country or at least a broad sub-national region. But two problems arise here. First, the taxes may have to be very high, especially if they are uniform, since such an approach fails to take account of heterogeneous soils and topography, as well as differences in location relative to receiving waters[9]. But if taxes are not uniform, the spectre of "smuggling" looms large, and the character of the necessary monitoring takes a big leap on the intrusiveness scale. This, then, is the background to the hunt for cheaper, less threatening and less contentious ways to take advantage of the opportunity presented by more or less unregulated NPS pollution, or, if one prefers, to begin to get a grip on something that has so far defied management efforts.

It is easy to see an analogy here that may have a powerful hold on some imaginations. The acid rain problem in the United States was only tackled in a serious way when the argument began to be believed that tradable permits to emit SO_2, and eventually NO_X would significantly reduce the cost of meeting whatever reduction target was ultimately chosen as policy goal. This belief rested, in turn, on some success with the earlier experiments with air pollution bubbles, offsets and credit banking, and on apparently quite substantial success with the trading of allowable lead additions to gasoline during the phase out of leaded gasoline. Would it not be outstanding to find that some form of economic implementation instrument could make the NPS problem solvable politically? Particularly enticing, at least in the OECD context, is the possibility that the direction of trading would be

---

[7] For an overview of the BMPs employed in the US, see Cestti et al. (2003) and USEPA (2003a).

[8] Such systems exist in Sweden and The Netherlands, while Denmark imposes a tax on household use of nitrogen fertiliser.

[9] See, for example, Lintner and Weersink (1999).

such that farmers would be paid by PSs to control runoff rather than being forced to bear the costs. Certainly, the United States Environmental Protection Agency (USEPA) seems to buy into this vision, for it has twice endorsed water quality trading as a potential solution to lingering NPS pollution problems – in 1996 (USEPA 1996a; USEPA 1996b) and again in 2003 (USEPA 2003d). Indeed, such schemes are currently operating in a number of basins, with perhaps the best-known occurring in the Tar-Pamlico basin in North Carolina[10].

Now, why is all this background from the OECD world, and mainly the United States at that, relevant to the developing world? Principally, we maintain, because the NPS problem is almost certain to grow in developing nations, both absolutely and relative to PS pollution, and to do so for the same mix of reasons outlined above. Thus, in Figures 2.4-2.10, we see evidence of an industrializing agricultural sector, certainly in Asia (Figure 2.4) and, on a different scale, in South America (Figure 2.5). In Figure 2.6, the increasing importance of developing country fertiliser use overall becomes evident. For a closer look at two Latin American agricultural powers, Brazil and Argentina, consider Figures 2.7, 2.8, and 2.9. In Figure 2.7, we see how quickly soybean production has been increasing in the two nations; and in Figure 2.8, we see the impressive gains in soybean yields per hectare relative to those obtained in the United States. In Figure 2.9, the differences in maize yields suggest that even greater fertiliser use is on the horizon as the countries play catch up on the yield scale[11]. Finally, Figure 2.10 illustrates another aspect to the rapid growth in grain production in South America – its availability as a cheap source of animal feed. The United States experience has shown that modern intensive livestock production practices can have a disproportionately large impact on nutrient runoff, as is shown for the Chesapeake Bay watershed in Table 2.2. Further, across the developing world, the same disincentives to regulatory efforts exist: lack of an ability to monitor, the resulting perception that more intrusive approaches will be necessary, and the political power of agriculture, motivated to action by the fear of intrusive regulation[12].

---

[10] There are also experiments going on with "offset" type programmes. These do not involve caps applying to all sources, but rather requirements for (usually) point source reductions more severe than required under the technology-based effluent standards, along with the introduction of possibilities for buying part of the extra reductions from unregulated NPSs. Examples are described in Woodward and Kaiser (2002); Rousseau (2001); Environomics (1999).

[11] See also Schnepf et al. (2001).

[12] Just by way of supporting examples here, detailed country studies, done for an Inter-American Development Bank (IDB) workshop in 2003, of Brazilian and Mexican uses of economic incentives in the control of water pollution and of withdrawals, reveal that in neither country does agriculture face any sort of effluent charge, and the withdrawal charges levied on irrigation water are trivial compared to those on domestic and industrial users. (Seroa da Motta and Feres 2003; Saade Hazin and Saade Hazin 2003)

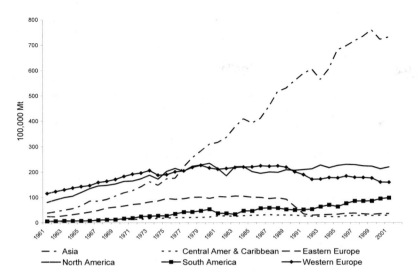

Fig. 2.4. Fertiliser Consumption trends from around the world (Source: FAOStat Agricultural Data, 2004)

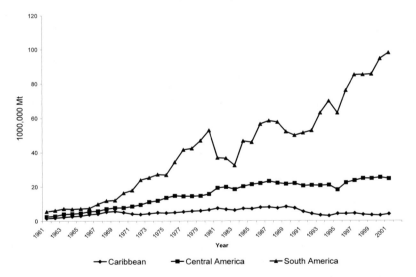

Fig. 2.5. Fertiliser consumption trends in Latin American and the Caribbean (Source: FAOStat Agricultural Data, 2004)

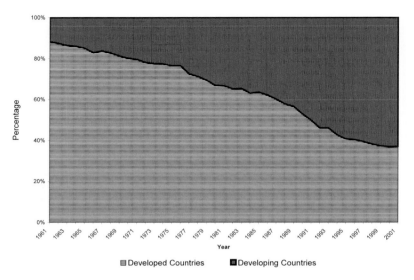

Fig. 2.6. Developed countries' decreasing share of world fertiliser consumption (Source: FAOStat Agricultural Data, 2004)

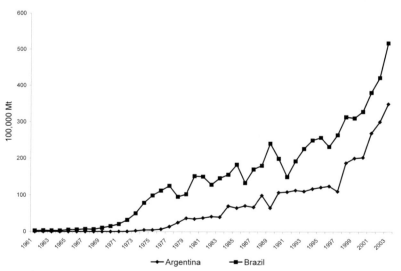

Fig. 2.7. Soybean production in Argentina and Brazil (Source: FAOStat Agricultural Data, 2004)

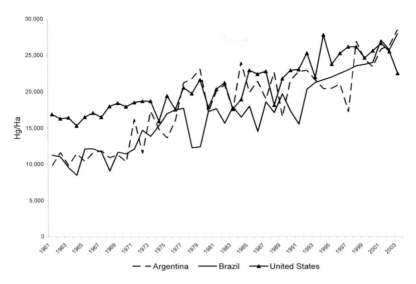

Fig. 2.8. Trends in soybean yields in Argentina, Brazil and the United States (Source: FAOStat Agricultural Data, 2004)

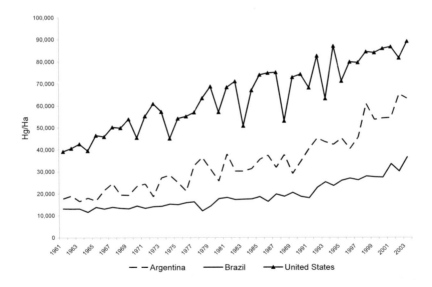

Fig. 2.9. Trend in maize yields in Argentina, Brazil and the United States (Source: FAOStat Agricultural Data, 2004)

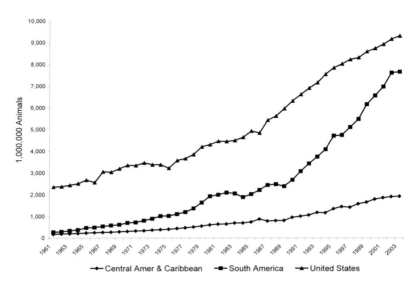

Fig. 2.10. Poultry meat production in Latin American and the United States (Source: FAOStat Agricultural Data, 2004)

Table 2.2. Average Phosphorous and nitrogen loads by land use in the Chesapeake Bay

	Total Phosphorous Loads	Total Nitrogen Loads
Land Use	(kg/ha)	(kg/ha)
Forest	0.1	4.3
Pasture	0.4	7.9
Livestock operations	460.0	2,302.0
Conventional tillage	2.6	25.2
Conservation tillage	2.0	20.5
Hay	1.7	11.0
Urban areas business and residential	0.9	11.1
Atmospheric loads	0.7	16.2

Source: Cestti et al., 2003.

In this chapter, we shall argue that, even though there are ways to introduce economic incentives, operating at the margin, to the NPS problem, the basis for doing so is hugely information intensive. It requires that extensive knowledge of local soils, topography, cropping patterns, and weather be combined with sophisti-cated models capable of predicting, with acceptable accuracy and precision, the implications for NPS discharges of specific BMPs on specific types of farms, with different soils, and in different climatic regions. Such data, and practical opera-tional versions of such models, are by no means widely available even in the rich-est nations. Thus, setting off on a policy experiment in developing countries that will require their availability from the beginning seems quixotic. It may be possible to reduce the political problem by spreading the financial burden of on-farm NPS reductions beyond the boundaries of the farm via non-marginal subsidy payments to farmers. And, if these subsidies are paired with incentives to promote information gathering, it may also be possible to serve the long run interest in better knowledge and therefore better tools. But, there will be no free lunch politically. Taking real advantage of the lower costs of nutrient control via NPS control will require bringing farms within the water pollution control system in an involuntary way.

A bit more perspective on how one OECD country, the United States, got where it is today relative to NPS pollution control, and how this seems to be driving efforts to introduce a system of tradable discharge "rights," may be useful. For one thing it will point up the costs of leaving whole categories of pollution sources out of initial efforts to protect and improve ambient water quality. For another, it will lead us back to the central difficulty – the lack of ability to monitor discharges – and thus, will force us to consider ways around that obstacle.

2.2 United States Experience

To keep the discussion reasonably short, we begin in 1972 when Congress overhauled the CWA with the Federal Water Pollution Control Act (FWPCA). The FWPCA was, in large part, a response to the shortcomings of the existing policy of state-based ambient water quality regulation, and it represented the first serious effort by the federal government to control water pollution in the United States. The FWPCA fundamentally altered water quality regulation in the United States by consolidating power for regulating water quality at the federal level and by imposing emission standards on individual polluters[13]. Thus, while the objective of the CWA is an ambient one – to "restore and maintain the chemical, physical, and biological integrity of the nation's water" – the primary tool for achieving this objective, at least since the 1972 amendments, has been a technology-based permitting programme for point sources ambitiously titled the National Pollution Discharge Elimination System (NPDES)[14]. The actual responsibility for administering the programme has, for the most part, been devolved to the states, although USEPA retains substantial oversight of the state programmes (Gallagher 2003: 279-81). NPSs, on the other hand, are exempt from NPDES permitting requirements. Instead, Section 319 of the CWA imposes on states a series of planning and reporting requirements for reducing NPS pollution, but includes no penalty for those that fail to meet these requirements (Title 33: Navigation and Navigable Wa-

---

[13] See Houck (2002) for a more detailed history of the FWPCA.

[14] The CWA defines a "point source" as "any discernable, confined and discrete conveyance," and lists a number of examples, but also specifically excludes "agricultural stormwater discharges and return flows from irrigated agriculture" from the definition. 33 U.S.C. §1362(14).

ters, Chapter 26: Water Pollution Prevention and Control, Subchapter III: Standards and Enforcement, Section 1329: Nonpoint Source Management Programs, United States Code). States, in turn, have responded to the challenge of regulating NPS pollution with a variety of initiatives, including land use controls for sensitive areas, such as land along streams or near wells[15]. Individual initiatives aside, the principal legacy of Section 319 has been a series of voluntary programmes for agricultural NPS – primarily education, technical assistance and cost shares for agricultural BMPs (Malik et al. 1994) – largely funded by federal grant programmes.

Thus, the CWA effectively limits the possible responses to the growing relative importance of NPS pollution. However, there are three efforts worth mentioning. First, the distinction between point and nonpoint sources is increasingly being blurred. For example, in 1987, Congress amended the CWA to regulate storm water discharges from industrial activities and municipal storm sewer systems serving more than 100,000 people (Gallagher 2003:300-302). USEPA has also recently got into the act, by broadening the regulatory definition of concentrated animal feeding operations ("CAFOs")[16]. Second, the FWPCA retained at least one vestige of the old state-based ambient water quality standards and this programme has become the focus of NPS control efforts. The CWA requires that each state designate intended uses for all of its waters and establish ambient water quality standards consistent with those classifications (Gallagher 2003:292-293). These standards serve a number of purposes. For one thing, PS dischargers that have "a reasonable potential to cause or contribute to" a violation of these standards are subject not only to a technology-based emissions standard, but also a potentially more stringent water quality-based standard (Gallagher 2003:293-294). Also, Section 303(d) of the CWA requires that states identify waters that fail to meet water quality standards after point sources have complied with NPDES permit requirements, calculate the total maximum daily load ("TMDL") of pollutants that these waters can receive and still be able to meet the applicable standards, and, finally, allocate these loads among dischargers. Section 303(d) was largely ignored by both USEPA and the states until a series of lawsuits resulted in court-ordered listings of impaired waters and schedules for TMDLs (Houck 2002:5).   Since TMDLs are not explicitly limited to PSs, they have been championed as the tool to finally regulate NPS pollution. However, the legislative history and wording of the statute make it clear that Section 303(d) was largely an afterthought and provides, at best, a narrow foundation for much in the way of regulatory structure, especially one that attempts to expand the traditional, if not intended, reach of the Act[17]. As a result, the future shape and direction of the TMDL programme, includ-

---

[15] For example, at least four states - Georgia, Oregon, Massachusetts and Wisconsin - have adopted comprehensive statewide buffer regulations (Environmental Defense 2003:4). Maine has a similar system, but one that does not touch existing agricultural operations.

[16] CAFOs were explicitly included in the definition of point sources from the beginning, but EPA recently lowered the threshold for triggering the CAFO designation and revised the permitting requirements applicable to CAFOs (USEPA 2003c).

[17] For a more optimistic view of the potential of Section 303(d) for regulating NPS pollution, see Tobin (2003).

ing whether it will serve as an effective basis for NPS regulation of any kind, are far from certain.

Third, as part of the Clinton Administration's programme of reinventing government, USEPA issued a brief policy statement supporting the notion of "water quality trading" in January of 1996 (USEPA 1996b) and supplemented the statement with a more detailed "Draft Framework" document in May of the same year (USEPA 1996a). USEPA's endorsement of trading in 1996 was limited in a number of important ways. Most notably, USEPA opted for an "offset" programme over a true watershed-based "cap and trade" system. The difference is that in an offset programme, a polluter under regulatory pressure to make additional reductions, such as a PS confronted with more stringent water-quality based effluent limitations, would be allowed to purchase the required reductions, or offsets, from other sources in the watershed, while in a cap and trade system a maximum quantity of a particular pollutant for a particular watershed would be determined, and tradable rights to emit, in aggregate, up to this level of pollutant would be distributed to dischargers, point and nonpoint, within the watershed. Although the documents specifically enumerate a number of different possible trading scenarios, it is clear that the greatest potential lies with the purchase of offsets by PSs from NPSs[18]. In January of 2003, USEPA reaffirmed its support of offset trading in a new water quality trading policy statement (USEPA 2003d)[19]. While it appears to evidence a somewhat stronger commitment to trading, the new policy statement specifically limits the endorsement of trading to programmes for nutrients, sediments and "cross-pollutant trading for oxygen-related pollutants." (USEPA 2003d: 4)[20].

It seems clear that the USEPA is labouring under some serious handicaps created by the lack of legislative authority. In particular, the effort to introduce trading, which may be seen as a subsidy programme funded not by the government but by PSs, depends on TMDLs and water-quality based effluent limitations for its force (Bartfeld 1993: 71-4). But the problem of policing the trades is, in effect, finessed. The suggestion has been made by commentators on the evolution of NPS policy that trades will be enforced by the PS buyers (Boyd et al. 2004), but this seems to us to imply the creation of potentially very high transaction costs – quite possibly high enough to discourage trading entirely.

---

[18] "An analysis of President Clinton's 1994 Clean Water Initiative (CWI) suggested that nutrient trading could lower the cost of implementing the initiative by $658 million to $7.5 billion, based on total incremental costs estimated to range from $5 billion to $9.6 billion. The bulk of these savings (75 to 92 percent) would be realized from point-nonpoint trading, while the remaining savings were expected to come from point-point trading (USEPA 1994)." (Faeth 2000: 16-17).

[19] USEPA's Region 10 and more recently USEPA's Office of Water have published similar water quality trading assessment handbooks that provide detailed guidance on evaluating a specific watershed's potential to benefit from a water quality trading programme (USEPA 2003b; USEPA 2004).

[20] The 1996 Draft Framework only provided that "[c]areful consideration [be] given to the types of pollutants traded." (USEPA 1996a: 2-4).

By way of follow-up to this last point, in the next (third) section we shall present some observations about the NPS problem setting. Some of these points have already been mentioned briefly above but are here brought together in a way that we hope will emphasise the central place of the monitoring difficulty. On that basis, in the fourth section, we shall describe what we take to be the key role of predictive modelling in any effort that seeks goals for NPS control defined either in terms of discharge reductions or of ambient quality improvements. We shall also spend a little time making the notion of the required data and models more explicit, and consider the extent to which practical and widely applicable versions of such models are available. The final section will be devoted to a suggestion for an NPS strategy for developing countries. It will build both on the United States experience and on our general arguments and will begin with requiring changes in some farm practices, such as the adoption of particular BMPs, seen as policy goals in themselves. This will likely be a conceptually unsatisfactory outcome for economists but may nonetheless provide a practical starting point, and a lever with which to pry loose information necessary for more sophisticated efforts.

2.3 Some Observations

Researchers analysing the NPS pollution problem sometimes seem to miss the proverbial forest for several technically enticing trees inherent in the subject. For example, it is common for NPS papers to stress the large number of individual sources and the stochasticity of their discharges (because of the dependence on rainfall events) when discussing difficulties for regulation, but to make very little of the monitoring problem (Braden and Segerson 1993). And when it comes to suggesting solutions, the tendency is to stress the economist's favourite criterion, static economic efficiency, as the basis for judging the performance of the alternatives. In this section, we introduce seven observations intended to bring the forest into better focus and to put in play additional bases on which to compare alternative solution ideas.

1. By far the single greatest problem posed by NPS pollution is that we cannot now measure the actual discharges of individual economic units, whether farms, forests, or golf courses. We can measure the ambient quality results produced jointly by the several (or many) sources in a sub-basin. But only when quality at a point on a stream is influenced by one, and only one, source can we infer a single source's discharge, and even in that highly artificial case, the inference might well be uncertain because of surface-ground water links and nutrient deposition from precipitation.
2. If you cannot measure it, you cannot charge for it or trade it, at least not in the straightforward way assumed for standard discussions of emission or effluent charges and tradable discharge permits. In particular, the by-now-famous cap and trade system, introduced in the United States in the Clean Air Act of 1990 as a central part of the acid rain control effort, depends heavily for its success on continuous discharge monitoring (Bell and Russell 2002).

3. Liability approaches naturally suggest themselves when only indirect measures of relevant actions are available, because liability penalties rest on results after the unobserved actions have been taken (or not) and, in effect, worked their way through some more or less complex natural or human (or combined) system. Indeed Segerson (1988) suggested what amounts to a joint liability system for NPS pollution control. Much more recently Romstad (2003) enhanced the idea with rules for opting out of what becomes a "team" effort at control. The practical problem here is that when several parties jointly cause the result that is observable, and their contributions are not separable, as they are not in the NPS context, each must, for static efficiency, be liable for the full marginal damage attached to the result. This implies that in general there will be substantial overpayment in total, and only by some arbitrary redistribution formula, such as that suggested by Hansen (1998) and Horan et al. (1998) can the total costs of the parties be made to add up to the actual total damage, while preserving the appropriate incentives at the margin. This combination is likely to prove extremely unpalatable politically[21].

4. Other commonly suggested lines of attack are indirect, operating through inputs and the production process itself (Ribaudo et al. 2001; Lintner and Weersink 1999; Shortle et al. 1998; Huang and LeBlanc 1994; Braden and Segerson 1993). Thus, continuing a focus on nutrient problems, farmers' purchases of commercial fertilisers offer a taxation target; and their application schedules, procedures, and "dosing" rules of thumb can, in principle, be influenced by regulations, even perhaps by economic incentives[22]. Further, how farmers arrange and enclose their arable fields, how/if they till those fields, and how they rotate crops over multi-year periods all influence the nutrient loads their farms discharge (Sharpley et al. 1999; Carpenter et al. 1998). Each of these choices is open to influence by payment schedules or regulations. Thus, for example, the maintenance of natural barriers between cropped fields and watercourses could be made worthwhile by suitable subsides, or their absence could be taxed. No-till cropping could be encouraged in analogous ways. There are, however, two sorts of problems with these indirect instruments:

- First, indirect approaches suffer technically for their indirection. Most obviously, nutrient quantity purchased is not the same as quantity discharged; it is not even the same in general as a constant $(1 + x)$ times the quantity discharged (Braden and Segerson 1993); though sometimes, for simplicity, such an assumption is made (Claassen and Horan 2001). The amount of fertiliser running off from a farm, forest, or golf course will depend on how and when the purchased quantity is applied and the fine-grained details of topography,

---

[21] The discussion here assumes, as do the liability suggestions in the literature, that marginal and total damages in money terms are known or at least knowable, hardly a trivial assumption. For a more general discussion of the application of liability rules to NPS pollution see Ribaudo et al. (1999).

[22] For example, one can imagine tradable rights to apply particular amounts of purchased chemicals (Shortle and Abler 1997; Taylor 1975) or "home-grown" manure per acre per growing season (Jones and D'Souza 2001; Willett and Mitchell 2001).

soils, and cropping on the unit in question. Most obviously, this means that if the policy goal is to reduce discharges, and if the instrument is indirectly connected to that goal, imposing a uniform instrument value may be a very ineffective way to proceed. Also problematic is that the side effects of any single approach may well be undesirable. For example, restrictions on fertiliser application rates may encourage extension of cropping to less efficient acres, with counterproductive implications for soil and nutrient runoff. And encouraging no-till agriculture implies increasing the use of herbicides, which can cause their own environmental damages.

- Second, because any sub region is embedded in a larger market setting, regional, national, and even international, effects of basin-by-basin interventions can also play out via pecuniary externalities. Damping fertiliser use in one place can make it balloon in another because of crop price changes, unless the system applies uniformly across the nation. But to design a national system, where the waterways are hardly uniform in current condition, and thus in their "need" for NPS control, creates a special kind of design problem. The implication seems to be that even to get final overall results that are desirable, forget optimal, it will be necessary to solve a large, complex programming problem along the lines of that used in (Ribaudo et al. 2001).

5. On the encouraging side of the ledger is the recognition that the stochasticity of NPS pollution need not be seen as implying huge difficulties for policy formulation. Consider the following:

- All pollution discharges have stochastic elements, whatever their sources. These result from such causes as random equipment and human failures, and weather, as it affects demand for products, availability of inputs, and performance of treatment units. In most discussions of policy formulation and instrument choice for point sources these variations are ignored, or are accommodated by design elements. For example, in the United States SO_2 trading system, time is allowed at the close of each year during which sources may, without penalty, acquire allowances that "validate" their actual discharges during the year to the extent these were different than their "expected" discharges...those for which they had already purchased or owned allowances. (See Innes, 2003, for a theoretical examination of how intertemporal permit banking can be used to accommodate stochasticity of discharges).

- In the case of NPS nutrient discharge, the effect of the variance created by rainfall/runoff events tends to be damped by the way the nutrients interact with and affect natural aquatic systems. They, in themselves, are not toxic, nor do they deplete oxygen immediately. When suspended in the water column, they promote the growth of aquatic plants, particularly algae. Heavily fertilized water bodies will suffer from algal blooms under the right combinations of temperature and sunlight. When these die off, oxygen is depleted, and the depletion can be a serious threat to fish and other aquatic life. But the problem may happen days, weeks or months after the runoff event and possibly quite far downstream. The bloom events will themselves be stochastic, but their distributions are unlikely to be closely tied to the distribution of runoff events.

Further, because nutrients can settle along with soil particles and algal cells, in areas with slower flows, and subsequently become resuspended during storm events, the entire system can be seen as having a long period variability, with things generally getting worse as inputs continue. All this implies that policies can be aimed at average results over longer periods than single storm events. That is to say, the policy goals for runoff discharges can look quite similar to those for pollutants with chronic, as opposed to acute, effects[23].

6. There is what might be called an unfortunate obsession in the policy instruments literature, PS as well as NPS oriented, with static efficiency as the measure of the value of a policy implementation choice. While lower costs for the same environmental effect are certainly better than higher ones, obtaining the lowest costs (the statically efficient result) in a regional problem setting, where individual location matters, is in general a difficult problem. This is because each source, will, in general, have to be dealt with differently to take account of its unique location relative to the places at which the chosen ambient quality standards will apply. Finding the individualised instrument values requires constructing and solving a regional optimisation model that contains, at the least, both the discharge reduction cost functions for each source and a sub model of the natural world system that will transform the discharges into ambient quality levels (Bohm and Russell 1985)[24]. As already noted, for agriculture, the model will also have to take account of any links among the policy instrument to be used and input, crop and tillage choices. Further there is no general possibility of simplification via an appeal to a second-best result. That is, a non-optimal economic incentive instrument is not automatically cheaper than a regulatory approach (Russell 1987). And, in any event, static means just that. As soon as the world changes, which it is in the habit of doing constantly, the statically optimal solution, whether a set of optimal fertiliser taxes, or a set of application rules, will also change. The model will have to be solved again and a new set of instrument values promulgated if the efficiency is to be maintained.

7. But, one might object, is not the SO_2 cap and trade system for acid rain control seen as a success exactly because it lowered the costs of meeting the discharge reduction goal below the level predicted during earlier debates about that goal? Can we not then reasonably expect a similar outcome from application of more or less any economic instrument to the NPS problem? If we abandon the notion of trying to meet ambient quality targets and focus on the total quantity of pollution discharged in the region, we then wash one aspect of location out of the

---

[23] Thus, the Mississippi River carries very large quantities of nutrients into the Gulf of Mexico, and the long -term result seems to be a chronic "dead" ("hypoxic") zone seaward from the river's mouths. Policies being examined for this problem focus on annual average discharges (Ribaudo et al. 2001).

[24] It is more than a little surprising, given how long this result has been known, how often papers in the instruments literature open with some version of a throw-away line implicitly claiming that achieving static efficiency, when ambient quality is the policy goal, is just a matter of applying a uniform charge or creating a tradable discharge permit system.

problem. Still present is the location of each farm relative to the water body being polluted[25]. If we are willing to assume that complication away as well, then whatever discharge total we want to achieve will be achieved at least cost if we use a policy instrument that equalises the marginal costs of discharge reduction efforts across all of the sources in the region. This can be achieved through a tradable discharge permit system. The SO_2 system did just that...it involved a cap on total discharges, a cap that was disaggregated first to the level of the states covered in the language of the act and then to the individual large power plant sources to be limited in their discharges. Since the disaggregation was based on historical discharges, not on any knowledge of marginal costs, it would have been surprising to find that trading among the sources would not have been able to lower the total cost of meeting the cap. It helped as well that the covered plants were of different vintages, had different capabilities on the fuel-mix dimension, and likely also were in different situations with respect to long run fuel contracts. All this implied that marginal costs were likely to differ across plants at any arbitrarily chosen allocation of allowable discharges[26]. Thus, while it might be true in a particular real case that a discharge cap and trade system for NPS pollution control would be cheaper than the same cap alone, assuming for the moment it could be made to work in the absence of monitoring capability, it is by no means a certainty, depending as it does on the details of the application. In any event, it seems important for the long run to examine with some care the options for getting around or relaxing the monitoring constraint. At the same time, it will be useful to look at the incentives provided for information revelation and the development of environment-saving technology, including better monitoring technology as well as agricultural technology that would allow lower application rates and improved retention of what is applied in the field[27].

---

[25] Lintner and Weersink (1999).

[26] This situation may be contrasted with that obtaining in the Fox River when an experiment with tradable water pollution permits was undertaken (David 2003). This also involved a single industry, pulp and paper production. But the marginal costs of the covered sources tended to be quite similar, in part because paper mills tend to use a variant of the technology used in municipal sewage treatment, because the mills in the basin had added the treatment works at about the same time, and because there are not even strong economies of scale for this technology because greater treatment flow rates for larger mills are achieved by replicating units rather than building bigger and bigger units. In the experiment no trading took place, so no cost savings could be claimed. There were other flaws in the design as well, possibly even more important than the lack of marginal cost differences, such as the stipulation of onerous review requirements for any proposed trade. Perhaps most astonishing, the rules did not allow that trades could be justified by cost savings. The overall lesson is that, monitoring difficulties aside, there is nothing automatic about even cost savings, let alone least cost attainment, in the pollution control policy business.

[27] "The problem is that about one half of every metric ton of fertiliser applied to fields never even makes it into plant tissue but ends up evaporating or being washed into local watercourses. A combination of better timing of fertiliser applications, more exact calcu-

2.4 Dealing with the Monitoring Problem

To this point, the argument has stressed that our inability to monitor individual source discharges rules out attacking NPS pollution in the straightforward ways that point source pollution can be managed, be they regulatory or incentive based. Further, the apparently easy ways around the monitoring problem, via input taxes or restrictions or the use of a version of joint liability, turn out not to be so easy in practice, because of large information needs, complex secondary effects, and intrusive (if different) monitoring requirements; and, in the liability case, because of the prospective overpayment problem. There are, however, two other approaches worth considering, for they have the ability to shift the burden of proof to the sources from the agency, thus relaxing the iron grip of inadequate (or excessively intrusive) monitoring technology. One of these approaches requires subsidising source discharge reductions from an agreed-on base, challenging the source to prove that the reductions are real. The second is roughly the opposite financially: assuming the worst...discharges at the base level...and charging accordingly, again challenging the source to prove better performance to qualify for a lower charge, or what has been referred to as a presumptive charge (Eskeland and Devarajan 1996). The key to either approach is the adoption of a publicly available model, or set of linked models, that can be shown to predict, with satisfactory accuracy and precision, discharges from any individual property, for any set of assumptions about the relevant topography, soils, and weather, as well as the management practices and application rates and timing employed on the property. Each property owner can claim discharge reductions based on claimed changes in the underlying actions. The model will verify that, if the changes are real, the discharge reductions will be produced. The changes themselves will have to be verified on the ground, but the essence of the shift in the proof burden is that it is now in the owner's interest to invite the required monitoring even if it is intrusive.

The idea of using models in such a central way may sounds to some like a version of the infamous can opener of the classic economist joke. But to accept this judgment would be unfair. First, it is clear that the technology exists to create such models, though as we caution below, that is a long way from saying we could launch such a system in the near future, even in the United States. Second, and as important, such uses of *a priori* models are far from unknown in the pollution control policy world. For example, regional airshed models have long been used in the definition and approval of proposed "offset" transactions between air pollution sources. And trades under the RECLAIM air pollutant trading scheme in Southern California make use of trading ratios that amount to summary versions of the results of model runs. At the theoretical end of the policy scale, the notion of tradable ambient quality (impairment) permits, shown 30 years ago by Montgomery (1972), to be capable of producing the statically efficient arrangement of discharges in a region in a decentralized market setting, rests on an agreed use of a

lation of doses, and more accurate delivery could cut this waste substantially." (WRI 1998).

regional airshed model. This is because only source discharges, not specific source contributions to ambient quality degradation at points in the environment, can be monitored, and the allowable discharge rate for each source is that consistent with the portfolio of ambient permits owned by it. The consistency is demonstrated by running the model.

The symmetry of subsidies and charges at the margin is well known. The use of marginal subsidies to affect behaviour in situations where actions are difficult or impossible to monitor is familiar, though the familiarity may be concealed behind the name we give the common systems: deposit-refund, or DR for short. DR as a policy tool is used to encourage the proper disposal of items or substances that are so easy to dispose of improperly that forbidding or threatening to punish such actions amounts to an empty gesture. Examples are empty beverage containers, worn out auto or boat batteries, and used engine lubricating oil. Their use has also been suggested for toxic chemicals that are not dissipated in use, such as cleaning solvents (Russell 1988) and pesticide containers (Malik et al. 1994).

Unfortunately, the self-financing characteristic of the DR approach does not seem capable of being captured in the NPS context, because the essence of the use of fertilisers and pesticides is their dissipation in the environment. Use equals disposal, in short. This is not to say that marginal subsidies cannot be used, only that, if they are, their financing must be faced and will present a political problem to be overcome. For example, fertilisers could be taxed, with the proceeds earmarked to pay the subsidies to those who qualify for them. Alternatively, since farmers are already heavily subsidised out of general revenues in most countries, that could be the source of the subsidy payments. In part because of this revenue problem, and in part because subsidies have potentially perverse effects away from the margin (by encouraging entry/discouraging exit through their income effects) the presumptive charge will hold considerable interest for many policy makers. Notice, finally, that once we accept the applicability of predictive modelling as the basis for charges or subsidies, we have also opened the door to the use of tradable discharge permits in combination with a basin or sub basin "cap" on total discharges, translated into individual permit allocations. Proposed trades must be "verified" using the same combination of modelling of the claimed actions and monitoring of the actions on the ground as applications for subsidies or for relief from the presumptive charge level.

A few additional comments are in order at this point. First, the definition of the "base case," from which reductions qualifying for a subsidy are measured and to which the presumptive charge applies, will be contentious and involve a sort of boot-strapping problem. That is, since by definition no measurements of discharges will exist, they must be generated by the same model, using historic patterns of application levels, etc. But those historic patterns will likely not have been observed either, so there will be considerable scope for dispute about how to define the base. In this regard, the interests of the parties who are to be charged or subsidised will be opposite for the two systems. For a subsidy, or for a tradable permit scheme in which the original endowment of permits available to trade is free, a high baseline will be desirable. For the charge, a low one will be sought. In each case, the agency interest will be opposite to that of the farmers. It may be that there is so much data collected as a matter of course in the agricultural world of OECD countries that disputes can be settled with relevant facts. This seems much less likely in the developing world.

2.5 Notes on the Necessary Models

It may be useful, as background for some information that puts a bit of flesh on the bones of the assertions about model requirements in the previous section, to begin by contrasting the NPS situation with that of a representative PS. Thus, consider a municipal sewage treatment plant or a paper mill. Under any public effort to protect ambient water quality, both will face restrictions, in some form, on allowable discharge amounts of solids and of organic materials that deplete instream oxygen, as they are broken down to more stable forms. Both will very likely respond to the restrictions by installing similar technology designed to settle and filter out suspended solids (so-called primary treatment) and to take dissolved organics out of the solution with the aid of bacteria and chemicals (secondary treatment). A tertiary level may be added if extra nutrient or organic chemical reductions are required. But all these processes discharge through the same pipe(s) to the receiving water. To sanitary engineers the treatment processes are interesting challenges. Understanding them well enough to model, and then using the models to explore different designs and operating regimes in order to increase the cost-effectiveness of removal is a fundamental skill and a research area of some importance. But from the regulatory point of view what matters is what is measurable at the end of the discharge pipe. Establishing and, more importantly, monitoring compliance with limits on discharges (or charging for discharge units) does not depend on an understanding of, much less the frequent use of models of, the black boxes that constitute the treatment plants.

For the NPS regulatory exercise, however, the analogues to the treatment process engineering models are central. It is as though, in the municipal sewage treatment plant, we could not measure discharge directly, but could observe such details as household water demand, the extent and type of industrial and commercial contributors to the sewer system flows, and the length of that system. Or, for the paper mill, if we could only know the capacity of the mill, the pulping process, and the types and quantities of its finished paper products. We would have to "monitor" discharges indirectly, by modelling the town or the mill to obtain estimates of raw wastewater volumes and pollution loads, and then modelling the installed treatment plant to estimate the resulting discharges. This is the situation for the farm as pollution source. We have to know rather a lot about the land on which it sits, the way it operates, the weather pattern it faces, in order to estimate its "raw waste water loads"…what would run off the land in the expected value sense, under the known pattern. Nature, in effect, is a key part of the process by which raw waste loads are generated and provides the "treatment plant," the link between those raw loads and what actually enters the stream, though nature can be helped along by farmer actions, such as leaving verges of fields unplowed, or not plowing at all. We cannot measure the resulting discharges, so we must know enough to model the entire system if we want to have some confidence in our estimates of them.

If models are to be central to the management of NPS pollution, those models must be technically acceptable, which almost certainly means that they must rest on sound conceptual foundations; must pass what we might call *a priori* tests of reasonableness in terms of the inputs they can accept and the outputs they can produce; and must be validated by tests in which their predictions match, within some acceptable error bounds, the actual data from test fields, farms or watersheds. Consider each of these dimensions in turn:

- The requirement that the conceptual foundations be sound may be restated briefly as a requirement that no physical laws be violated and that processes generally agreed to matter in NPS pollution generation must be reflected in the mathematics.
- The list of relevant inputs that must be accepted by the models reflects the set of circumstance that may be expected to differ across the range of NPS generation situations[28]. Thus, soil types, field slopes, and crops grown will all have an effect on how much runoff is generated by a rainstorm of particular intensity and duration. Since the models must predict results for any place, they must be capable of accepting or generating internally, the parameters that describe local rainfall distributions, and do this by season, so that results will be sensitive to when the rain falls relative to the condition of the fields. In addition, of course, it must be possible to reflect in the model those choices at the heart of the policy option set: the actions the farmer may take to modify runoff results from a given storm, given his particular soils, slopes, and cropping. These may include: changing rotations, so that the order of crops is different, even if their identity is the same; changing tillage methods, especially from conventional to no-till; introducing streamside buffer strips; introducing or restoring formerly existing wetlands; and changing fertiliser application rates or, more subtly, application timing over the crop cycle. Notice that the details of how each such input is brought into the model may be quite complex, so that drawing a black box called the runoff model and showing beside it a list like the one above does not carry the modelling process very far, though it may help to make the point intended here.
- Outputs desired from a NPS model include volumes of water entering specific watercourses, the suspended solids load in that water, and at least the nitrogen (N) and phosphorus (P) concentrations as well. This last requirement is complicated by the fact that the form of the nutrients may vary with the form of fertiliser, and that each, but especially N, may come off the field in several

---

[28] We concentrate here on agriculture as a source type. In the world of NPS modelling, there are models that are specialized for urban and suburban situations. These will generally have different structures and reflect at least some different input data.

forms, with some of these forms tending to leach into the soil while others travel across the field[29].

- Determining the quality of the model's predictions is difficult for exactly the reason that the models are necessary...the inability to measure NPS pollution from particular farms. It will often be possible, however, to gather data, make model runs, and check the results against measurements for well-defined sub-basins, for which all the contributing farms or parts of farms can be character-ised as above, and located relative to each other and the water courses. Then, accurate prediction of the water quality of the receiving water, given the input data, will suggest that the model "works" as intended[30]. The more variety in the set of the sub-basins used as test beds, the more useful the model is likely to be.

So far so obvious, if not so easy. The practical problem for policy development and implementation is that, while many NPS models exist, there does not seem to be a standard model that can be taken off the shelf and used as a platform on which to construct basin and sub-basin specific models. This is because particular models tend to reflect their provenance...the problem of greatest interest to the builders. Thus, Thomas et al. (1998: 1), assert that, "...no model is designed to meet all the needs of students, researchers, extension agents, regulatory agencies, planning organizations, consultants and environmental groups. The majority of widely used water quality models were designed for particular applications (whether spatial or temporal)." Another author puts it more colourfully: "There are already so many environmental water quality models (EWQMs) that sentences could be made from their acronyms... Because of the multitudes of EWQMs, en-vironmental professionals have a daunting task choosing an appropriate model. Furthermore, most models seemingly have 'unreasonable' parameter and variable input requirements." Reyes (1998: 2) and Joubert et al. (1996: 1) go to the heart of the practical matter: "...with so many methods available, why is it so rare to find them actually used to support local land use decisions? Often they are either too complex for planning staff to use, or costly field monitoring is necessary to cali-brate the model. When a model is easy to apply, it is commonly applicable only to certain types of watersheds [or] soil conditions, or the results are so generalised as to be of little value."

Thus, there is no lack of models. If anything, there is such a bewildering array that deciding among them for the intensely practical work contemplated here would itself amount to a major research project. And, in the end, no single one might be best (or even acceptable) in all applications. This seems clearly to pose political, and quite likely legal, problems. Even if a standard "platform" can be found or developed, its successful application across substantially different re-

---

[29] We ignore here, for simplicity of discussion, the important matter of pesticide runoff, to which most of the same comments apply.

[30] Parson et al. (1998), provide an example of such a validation exercise for a particular NPS pollution model with data from a US watershed and Fistikoglu and Harmancioglu (2003), describe an example using data from Turkey and stress the problems caused by lack of basic data, likely to be the condition found in developing countries.

gions is likely to require what model builders euphemistically call "retuning," but which a lay person might be tempted to call tinkering with the supposedly fundamental structure of parameter values. This can itself be expensive and time consuming, and can produce a weak point for attack by those who believe they are, or will, suffer because of the model's use for predicting discharges.

2.6 Suggestion for an NPS Regulatory Strategy in the Developing Country Context

The executive summary of the above material may be written as follows:

- The United States experience with NPS pollution holds both negative and positive lessons for nations that see this problem as a current or future concern.
- The greatest negative lesson is that avoiding the problem because of its political and technical difficulty is a big mistake. The problem can only get worse and the interests more deeply entrenched.
- The major positive lesson is that in the long run, economic incentive approaches may, indeed, hold the promise of reducing the cost of maintaining ambient water quality, in particular by allowing sources of pollution with the lowest cost to do the most discharge reduction. But to tap into this promise, extensive and quite detailed information, embedded in flexible and robust natural systems models of the runoff generation process will be central.

On this basis, our suggestions for a NPS strategy for developing countries are as follows[31]:

1. Treat large, intensive animal operations as point sources.
2. Establish the principle of public control over NPS pollution (and thus the basis for eventual trading) via a general requirement that BMPs be put in place by every source. The legislation should include a list of BMPs that are, in effect, certified as useful. The implementing regulations could be more specific, matching the practices to the type of operation for which they would be considered sufficient for this stage. For example, for animal grazing operations, a certified BMP might be to fence the animals away from 95% of the stream banks, creating watering places every X thousand meters. For cropping, the requirement might be the creation of buffer strips between ploughed field margins and rills or streams. Where appropriate, wetland restoration or creation might be the

---

[31] We assume that, in developing countries as in the OECD world, NPS control will, in the long run, be a cheaper way to reduce N and P loadings than will tertiary wastewater treatment for cities and industry. In the short run, if there has been little pollution control effort generally, this need not be true.

BMP of choice. All this could be based on the existing literature's estimates of effectiveness[32].

3. The costs would be deductible for income tax purposes, if there are such taxes, or from land taxes, if those exist.

4. Farms could apply for further subsidies, but part of the subsidy agreement would be the opening of the subsidised unit to quite intensive monitoring to gather data on which to base the next stage of the process. This monitoring could include fertiliser purchases and manure generation or purchase, application timing and rates per ha, uptake by crops, uptake by buffer strips, loads in adjacent streams before and after introduction of the BMPs, evidence of decay of the effectiveness of the BMPs over time, and the cost of maintenance to hold decay at bay.

5. This could be organized as a giant experiment, with the subsidy/data-gathering exercises parcelled out in a systematic way across topographies, soil types (at least as estimated a priori if data are scarce), farm types and sizes, and climatic regions.

6. After some time, perhaps five years, perhaps more, an experiment in regulation could be launched on the basis of the knowledge gained. Under this, each source would be assigned a baseline permitted total discharge, inferred on the basis of the installed BMP and the new knowledge/modelling skill. If the NPSs were determined to be contributing to unacceptable ambient quality in a particular basin, a TMDL-type process would be triggered, and required reductions allocated across point and nonpoint sources. These would be tradable. If the allocation ignored location details, the importance of which is demonstrated in Lintner and Weersink (1999) that alone would open up NPS trading. But one would expect there to be PS/NPS trading as well in complex basins with urban as well as rural areas. The trades would involve the use of the models to infer required trading ratios for particular sources, and the basis of the seller's reduction would have to be specified, both for modelling and for monitoring purposes[33].

7. This experiment might be too big an enterprise to launch nationally all at once. But developing countries seem comfortable with basin-by-basin efforts, judging at least by the Brazil and Mexico case studies referred to earlier in the chapter. There are other complications as well. A major one may be the federal question…how to get cooperation from the states, especially the states in which agriculture is a major industry. But we believe one fairness precondition will be that in the initial phase every farm in every region/basin be required to install the BMPs, whatever they are determined to be.

---

[32] However, more options may translate into a more cost-effective policy. "As with other command-and-control type policies, there is no assurance that the controls required would be cost-effective, although maintaining farmer flexibility by offering a range of acceptable BMPs would improve cost-effectiveness." (Malik et al. 1994).

[33] There is also the possibility of presumptive charges, but these would be trickier, because of the tenuous linkage, through the farm's reaction to a particular charge, and the tracing of that reaction through the natural world to the resulting discharge.

Thus, the idea is to begin by asserting authority over NPSs, to use subsides as the lure to obtain access to information, and only later to attempt to take advantage of trading in discharge rights.

2.7 References

Adler RW, Landman JC, Cameron DM (1993) The Clean Water Act 20 Years Later. Island Press, Washington, DC

Bartfeld E (1993) Point-Nonpoint Source Trading: Looking Beyond the Potential Cost Savings. Environmental Law 23(1): 43-106

Bell RG, Russell CS (2002) Environmental Policy for Developing Countries. Issues in Science and Technology 69: 63-70

Bohm P, Russell CS (1985) Comparative Analysis of Alternative Policy Instruments. In: Kneese AV, Sweeney J (eds) Handbook of Natural Resource and Energy Economics vol. 1. Amsterdam: North-Holland pp 395-460

Boyd J, Shabman L, Stephenson K (2004) Water Quality Trading's Present and Future: Offsets, Watershed Permits, and Cap and Trade. Working paper

Braden J, Segerson K (1993) Information Problems in the Design of Nonpoint-Source Pollution Policy. In: Russell C, Shogren J (eds) Theory, Modeling and Experience in the Management of Nonpoint-source Pollution. Kluwer Academic Publishers, Norwell, MA

Carpenter SR, Caraco NF, Correll DL, Howarth RW, Sharpley AN, Smith VH (1998) Nonpoint Pollution of Surface Waters with Phosphorus and Nitrogen. Ecological Applications 8(3): 559-568

Cestti R, Srivastava J, Jung S (2003) Agriculture Nonpoint Source Pollution Control: Good Management Practices – The Chesapeake Bay Experience. World Bank Working Paper number 7, World Bank, Washington, DC

Claassen R, Horan RD (2001) Uniform and Non-uniform Second-best Input Taxes. Environmental and Resource Economics 19(1): 1-22

David E (2003) Marketable Water Pollution Permits as Economic Incentives: Point Source Trading In Wisconsin, Washington, DC. Inter-American Development Bank Technical Seminar on the Application of Economic Instruments in Water Management (February 27)

Environmental Defense (2003) Riparian Buffers: Common Sense Protection of North Carolina's Water. http://www.environmentaldefense.org/documents/2758_NCbuffers.pdf.

Environomics (1999) A Summary of US Effluent Trading and Offset Projects. Report prepared for Dr. Mahesh Podar. United States Environmental Protection Agency (November)

Eskeland GS, Devarajan S (1996) Taxing Bads by Taxing Goods: Pollution Control with Presumptive Charges. Directions in Development Series. The World Bank, Washington, DC

FAO (2004) FAOStat Agricultural Data, Food and Agriculture Organization. http://apps.fao.org/page/collections?subset=agriculture

Faeth P (2000) Fertile Ground: Nutrient Trading's Potential to Cost-effectively Improve Water Quality. World Resources Institute, Washington, DC

Fistikoglu O, Harmancioglu NB (2003) Integration of GIS with USLE in Assessment of Soil Erosion. Water Resources Management 16: 447-467

Gallagher LM (2003) Clean Water Act. In: Sullivan Thomas FP (ed) Environmental Law Handbook (Seventeen Edition). ABS Consulting, Rockville, Maryland, 271-339

Gollehon N, Caswell M, Ribaudo M, Kellogg R, Lander C, Letson D (2001) Confined Animal Production and Manure Nutrients. Agriculture Information Bulletin number 771. Resource Economics Division, Economic Research Service, US Department of Agriculture, Washington, DC

Hansen LG (1998) A damage based tax mechanism of nonpoint emissions. Environmental and Resource Economics 12(1): 99-112

Horan RD, Shortle JS, Abler DG (1998) Ambient taxes when polluters have multiple choices. Journal of Environmental Economics and Management 36(2): 186-199

Houck OA (2002) The Clean Water Act TMDL Programme: Law, Policy and Implementation. 2nd edn, Environmental Law Institute, Washington, DC

Huang W, LeBlanc M (1994) Market-based incentives for addressing nonpoint water quality problems: A residual nitrogen tax approach. Review of Agricultural Economics 16: 427-440

Innes R (2003) Stochastic Pollution, Costly Sanctions, and Optimality of Emission Permit Banking. Journal of Environmental Economics and Management 45(3): 546-568

Jones K, D'Souza G (2001) Trading Poultry Litter at the Watershed Level: A Goal Focusing Application. Agricultural Resources Economic Review 30(1): 56-65

Joubert L, Kellogg DQ, Gold A (1996) Watershed Nonpoint Assessment and Nutrient Loading Using the Geographic Information System-based MANAGE Method. Paper presented at Watershed 96. http://www.epa.gov/owow/watershed/Proceed/joubert.html

Lintner AM, Weersink A (1999) Endogenous Transport Coefficients: Implications for Improving Water Quality from Multi-contaminants in an Agricultural Watershed. Environmental and Resource Economics 14(2): 269-296

Malik AS, Larson BA, Ribaudo M (1994) Economic Incentives for Agricultural Nonpoint Source Pollution Control. Water Resources Bulletin 30(3): 471-480

Montgomery WE (1972) Markets in Licenses and Efficient Pollution Control Programmes. Journal of Economic Theory 5: 395-418

Newman A (1995) Water pollution point sources still significant in urban areas. Environmental Science and Technology 29: 114A

Parson SC, Hamlett JM, Robillard PD, Foster MA (1998) Determining the Decision-Making Risk from AGNPS Simulations. Transactions of the ASAE 41(6): 1679-1688

Reyes MR, (1998) Comparing the Inputs and Outputs of the GLEAMS, RUSLE, EPIC and WEPP Models. Paper presented at the Meeting of the American Society of Agricultural Engineers. http://www3.bae.ncsu.edu/s273/ModelProj/reyes98.pdf.

Ribaudo MO, Heimlich R, Claassen R, Peters M (2001) Least-cost Management of Nonpoint Source Pollution: Source Reduction vs. Interception Strategies for Controlling Nitrogen Loss in the Mississippi Basin. Ecological Economics 37(2): 183-197

Ribaudo MO, Horan RD, Smith ME (1999) Economics of Water Quality Protection from Nonpoint Sources: Theory and Practice. Agricultural Economic Report Number 782. Economic Research Service, United States Department of Agriculture, Washington, DC

Romstad E (2003) Team Approaches in Reducing Nonpoint Source Pollution. Ecological Economics 47(1): 71-78

Rousseau S (2001) Effluent trading to improve water quality: what do we know today? Working Paper Series. Faculty of Economics and Applied Economics Sciences, Center for Economic Studies, Energy, Transport & Environment. Katholieke Unviersiteit, Leuven, Belgium

Russell CS (1987) A Note on the Efficiency Ranking of Two Second-best Policy Instruments for Pollution Control. Journal of Environmental Economics and Management, 13(1): 13-7

Russell CS (1988) Economic Incentives in the Management of Hazardous Wastes. Columbia Journal of Environmental Law 13(2): 257-274

Saade HL, Saade HA (2003) Country Case: Mexico. Paper prepared for the IDB sponsored meeting: Water Charge Instruments for Environmental Management in Latin America: From Theoretical to Practical Issues, Washington, DC (February)

Schnepf RD, Dohlman E, Bolling C (2001) Agriculture in Brazil and Argentina: Developments and Prospects for Major Field Crops. Agriculture and Trade Report, WRS-01-3. Market and Trade Economics Division, Economic Research Service, US Department of Agriculture, (November)

Segerson K (1988) Uncertainty and Incentives for Nonpoint Pollution Control. Journal of Environmental Economics and Management 15(1): 87-98

Seroa Da Motta R, Feres JG (2003) Country Case: Brazil. Paper prepared for the IDB sponsored meeting: Water Charge Instruments for Environmental Management in Latin America: From Theoretical to Practical Issues, Washington, DC (February)

Sharpley AN, Daniel T, Sims T, Lemunyon J, Stevens R, Parry R (1999) Agricultural Phosphorus and Eutrophication. Agricultural Research Service (ARS-149), United States Department of Agriculture, Washington, DC (July)

Shortle JS, Abler DG (1997) Nonpoint Pollution. In: Folmer H, Teitenberg T (eds) International Yearbook of Environmental and Natural Resource Economics. Kluwer Academic Press, Dortrecht, The Netherlands

Shortle J, Horan RD, Abler DG (1998) Research Issues in Nonpoint Pollution Control. Environmental and Resource Economics, 11(3-4): 571-585

Taylor CR (1975) A Regional Market for Rights to Use Fertiliser as a Means of Achieving Water Quality Standards. Journal of Environmental Economics and Management 2(7): 17

Thomas DL, Evans RO, Shirmohammadi A, Engel BA (1998) Agricultural Nonpoint Source Water Quality Models: Their Use and Application. http://www3.bae.ncsu.edu/s273/ModelProj/thomas98.pdf

Tilman D, Cassman KG, Matson PA, Rosamond LN, Polasky S (2002) Agricultural sustainability and intensive production practices. Nature 418: 671-677

Tobin Erin (2003) Pronsolino v. Nastri: Are TMDLs for Nonpoint Sources the Key to Controlling the 'Unregulated' Half of Water Pollution? Environmental Law 33: 787-840

USDA (2003) Farms and Land in Farms: February 2003. National Agricultural Statistics Service, US Department of Agriculture, Washington, DC

USEPA (1989) Water Improvement Study. US Environmental Protection Agency, Washington DC

USEPA (1996a) Draft Framework for Watershed-Based Trading. US Environmental Protection Agency, Office of Water, EPA/800-R-96-001, Washington, DC. http://www.epa.gov/owow/watershed/framework.html

USEPA (1996b) Effluent Trading in Watersheds Policy Statement. http:www.epa.gov/owow/watershed/trading/tradetbl.html

USEPA (2001) Protecting the Nation's Waters Through NPDES Permits: A Strategic Plan. FY 2001 and Beyond. Office of Water, US Environmental Protection Agency, EPA-833-R-01-001, Washington, DC (June)

USEPA (2002) National Water Quality Inventory: 2000 Report. US Environmental Protection Agency, Washington, DC, Office of Water, EPA-841-R-02-001 (August)

USEPA (2003a) National Management Measures for the Control of Nonpoint Pollution from Agriculture. US Environmental Protection Agency, Office of Water, EPA-841-B-03-004, Washington, DC (July)

USEPA (2003b) Water Quality Trading Assessment Handbook: EPA Region 10's Guide to analyzing Your Watershed. US Environmental Protection Agency, Region 10, EPA 910-B-03-003 (July)

USEPA (2003c) National Pollutant Discharge Elimination System Permit Regulation and Effluent Limitation Guidelines and Standards for Concentrated Animal Feeding Operations (CAFOs). Final Rule. 68 Federal Register 29:7176-7274 (February 12)

USEPA (2003d) Water Quality Trading Policy. US Environmental Protection Agency, Washington, DC, Office of Water (January)

USEPA (2004) Water Quality Trading Assessment Handbook: Can Water Quality Trading Advance Your Watershed's Goals? US Environmental Protection Agency, Washington, DC, Office of Water (November)

Willett K, Mitchell DM (2001) An Integrated Model of Poultry Litter Permit Trading: A Firm-Level Analysis. Paper presented at Southern Economic Association Annual Meeting, Tampa, Florida

Woodward RT, Kaiser RA (2002) Market Structures for US Water Quality Trading. Review of Agricultural Economics 24: 366-383

WRI (1998) Nutrient Overload: Unbalancing the Global Nitrogen Cycle, World Resources 1998-99. World Resources Institute, Washington, DC

Role of Water Rights and Market Approaches to Water Quality Management

Sarah A. Cline, Mark W. Rosegrant and Claudia Ringler

3.1 Introduction

Compromised water quality threatens human health, the environment, industrial capacity, and agricultural production in many regions around the world. While the water treatment systems developed in most industrialised countries have curtailed many water related diseases caused by faecal contamination, important water quality problems still exist in these countries. Some of the main water quality concerns in developed countries today include trace chemicals and pharmaceuticals, as well as nonpoint sources of pollution from agriculture such as runoff from fertiliser, pesticides, and siltation (Davis and Hirji 2003). These nonpoint sources are more difficult to regulate than point sources and thus most countries have only recently begun to deal with them. In developing countries, on the other hand, water treatment and sanitation are still unavailable in many areas, leading to numerous water-related illnesses. Major water quality problems in developing countries include faecal contamination from untreated municipal wastewater, industrial effluents, and runoff from pesticides, fertilisers and herbicides from agriculture (Davis and Hirji 2003).

Water contamination can lead to serious human health effects. This is a particularly critical problem in areas where facilities for water and wastewater treatment are poor or non-existent. Water-borne diseases, primarily from the unsanitary disposal of human waste, cause illness and death, particularly among pre-school children. The UNDP-World Bank Water and Sanitation Programme estimates that approximately 6,000 people die from diarrhoeal diseases every day (Davis and Hirji 2003). The cholera outbreak in Peru and neighbouring countries in the early 1990s, which affected almost 400,000 people in 1991 alone, for example, was diffused by poor water supply conditions and overcrowding of the shanty towns that surround the coastal cities of the country.

According to the World Health Organization (WHO), by 2000, 85 percent of the population in the Latin American and Caribbean region had an improved water supply, while 78 percent had sanitation services. These percentages mask great inequity across countries and between urban and rural areas, however. While approximately 87 percent of the urban population has sanitation coverage, only 49 percent of those in rural areas are provided with coverage. Similarly, while 93 percent of the urban population has an improved water supply, only 62 percent of the rural population has comparable coverage. In general, the Caribbean countries tend to have a higher level of coverage than other countries in the region, with the exception of Haiti, which has less than 50 percent coverage for water supply and sanitation (WHO and UNICEF 2000).

In many countries, water pollution has traditionally been addressed using command-and-control approaches, which set standards to reach or maintain a certain level of water quality. While these approaches have often been successful for dealing with point source pollution, it is much more difficult to address nonpoint source pollution using such methods. This is particularly critical, as these sources contribute the largest amount of some pollutants in developing countries (Davis and Hirji 2003). For example, 75 percent of the water pollution in Colombia stems from agricultural nonpoint sources (Giugale et al. 2003).

In this chapter, we examine economic incentives, particularly market-based instruments used for water quality protection. In the first section, we discuss the various approaches. Next we provide examples of how different methods have been employed. Many of our examples are from the United States, where a variety of market-based approaches have been employed. However, general conclusions can be drawn from the successes and problems encountered in these programmes. We then discuss some of the constraints that may be encountered, particularly those facing Latin America and the Caribbean and other developing countries, and possible remedies. Finally, we propose steps for the future development of market-based approaches to water quality in the Americas.

3.2 Approaches

3.2.1 Water Rights

Water Rights are most often discussed in terms of water quantity issues and some system of water rights is found to operate in virtually any setting where water is scarce. The use and discussion of water rights in the context of water quality issues, however, has been mostly overlooked.

The market-based approaches to water quality issues discussed below are grounded, to a large extent, on the recognition of secure rights to water. The use and trading of effluent permits, for example, gives polluters a right to pollute a water source at a permitted level, which in turn can influence the availability of clean water available for other users. In fact, the idea of a "right to pollute" is a major concern that some environmentalists raise with market-based instruments.

Although water rights are not commonly used to deal explicitly with water quality, there is some scope for such use in areas where water rights have been defined for water quantity. For example, government authorities or conservation groups in the United States have, in some instances, purchased irrigation use rights from farmers in order to increase environmental flows. This could lead to decreased nonpoint source pollution from agricultural sources as well as increasing overall water quality through dilution of pollutants due to the increased environmental flows.

3.2.2 Market-Based Approaches

Market-based approaches to water quality and other environmental problems are often considered to increase cost-effectiveness and to provide incentives for technological innovation compared to the command and control approaches to environmental regulation traditionally used in many countries (Stavins 2000). The goal of these approaches is generally to reduce the environmental damage in question at the lowest possible social cost by aligning private and social costs. While many types of market-based instruments exist, some of the instruments considered for water pollution control include pollution charges, tradable pollution permits, and increasing the price of environmentally damaging inputs (by either taxing or removing subsidies) (Davis and Hirji 2003). Another innovative approach to enhance water quality is environmental service payments, where stakeholders interested in improved water quality pay for watershed conservation and management activities.

While market-based instruments for environmental protection are focused on attaining a certain level of environmental performance, their ability to raise revenue is also important in many regions. Historically in many Latin American and Caribbean countries, the primary role of these approaches has been to raise revenue (Huber et al. 1998). The revenue from market-based instruments can then be earmarked for programmes to reach certain environmental objectives. The earmarking of funds for environmental projects is controversial, however, and its success will depend upon the specific characteristics of each situation. Subsidising pollution reduction measures through earmarking of pollution taxes is not justified if the tax is set high enough to reach the set water quality goal, for example. The taxes are often set too low, however, and do not have the intended incentive effect (O'Connor 1999).

It should also be noted, however, that the administrative costs of set-up and implementation of these programmes can be as high or in some cases higher than those under command and control approaches. Many of the monitoring and enforcement costs involved with command and control regulations still exist for market-based instruments. However, additional costs may also be incurred due to design and institutional changes (Huber et al. 1998).

3.2.3 Pollution Charge Systems

Pollution charge systems assess a certain tax per amount of pollutant emitted by a given firm. Different firms will reduce pollution by varying amounts depending upon their marginal costs of abatement. This type of system ideally reaches a given level of pollution at the most efficient cost by allowing firms with high control costs to pollute more, while those firms with low control costs will pollute less. Effluent tax systems are also appealing because of their potential to promote pollution control innovations and their ability to generate revenues (Boyd 2003). The revenue-generating aspect can be particularly important to many regions since the investments required for water pollution control infrastructure are typically high.

The difficulty with this type of system is determining the appropriate level of tax to charge to obtain the most efficient level of pollution reductions. If charges are too low, pollution would continue at high rates and environmental objectives would not be achieved. On the other hand, if the charges assessed are too high, abatement levels would be very high but the revenues would be very low, which is a concern for those countries that wish to use these methods as a revenue-generating mechanism.

In addition, pollution tax systems are often hard to sell politically. Polluters are often against this type of regulation as they are responsible for the cost of implementing the control technologies as well as for taxes on uncontrolled emissions (Boyd 2003). Moreover, new firms are often held to higher standards than existing firms under such a system, leading to disincentives for entry of new firms. Environmentalists also often oppose pollution tax systems on ethical grounds since all firms are not required to abate the same amount.

Other difficulties include the fact that these systems are harder to implement than traditional command and control approaches. The possibility that the regulators who are setting the pollution charges could be influenced by involved parties raises additional concerns (Boyd 2003). Monitoring of sources also generates problems that become much more complex when nonpoint sources are considered. These sources are generally much smaller and harder to monitor than point sources. Due to these difficulties, taxes on inputs (such as pesticides and fertilisers in the case of agriculture) rather than on outputs may be more reasonable for nonpoint sources of pollution.

3.2.4 Tradable Water Pollution Permits

Pollution trading approaches to deal with certain types of pollutants have become fairly well established, particularly in the US. The first trading programme in the US was established in 1974 to allow limited exchange of emission reduction credits for five air pollutants (Kieser and Fang 2004). Additional trading programmes were created for leaded gasoline in the 1980s and sulphur dioxide in the 1990s. The first phase of the sulphur dioxide programme created by the 1990 Clean Air Act was quite successful, reducing SO_2 levels 30 percent below the target and generating cost savings of around $3 billion per year (Kieser and Fang 2004).

While these market approaches have become relatively accepted for air quality regulations, they have not traditionally been as widely used for water quality. The use of markets for water pollution has been gaining popularity in recent years, however, particularly in the US, where many tradable permit programmes have been developed since the mid-1990s.

Pollution trading programmes generally seek to achieve a certain level of environmental quality while minimising the abatement costs incurred by polluters. These programmes have appealed to policy makers in many areas not only as a means to decrease the costs of pollution reduction, but also to help meet current environmental standards that were not being met through traditional regulatory means. Woodward and Kaiser (2002) suggest five possible agency goals for a water quality trading programme: 1) reaching environmental goals defined by laws or regulations, 2) minimising the social costs of reaching a proposed environmental goal, 3) allowing the agency to maintain control over the programme while minimising legal risks and effort put into day-to-day programme operation, 4) minimising transaction costs by the participants, and 5) minimising the costs of initiating the programme for agencies and participants. Some authors suggest significant cost savings from water pollution or nutrient trading programmes. For example, a study in Michigan estimated costs of $2.90 per pound of phosphorous removed, while conventional regulations were estimated to cost around $24 per pound (Faeth 2000). As mentioned above, it should be noted that these programmes might also have additional administrative costs that can be significant.

When initiating a water quality trading programme, several legal issues must be considered with respect to the legal and institutional setting of the country where the regulations are to be implemented. First of all, it must be determined if such a programme is authorised under the current water quality regulations that are in force in a certain jurisdiction. The implementation of any trading programme must not violate current water quality regulations. It is also crucial that polluters monitor and report their emissions so that the agency with oversight authority will be able to determine if water quality standards are being met. In addition, as with other types of pollution trading programmes, there must be a legal entitlement for the pollution discharge. These entitlements must be transferable and enforceable in order for an effluent trading programme to work properly. Finally, the issue of enforcement is critical to ensure that the market functions effectively and that water quality standards are met (Woodward and Kaiser 2002; Woodward et al. 2002).

Pollution trading programmes may be established as closed or open systems. A closed system, often referred to as a "cap and trade" system, is an extension of traditional regulatory methods that sets a limit or "cap" on the amount of emissions allowed within a designated area before issuing permits to polluters. Polluters within the area are issued tradable emission permits for a designated amount of pollution, with the total allowed emissions not permitted to exceed the cap amount. This type of system is often used in areas where water quality goals are currently not being met. The less common open system is generally voluntary and uses existing regulations to assess a baseline level of pollution. Any reductions below the baseline create reduction credits that can be banked, traded, or used to comply with regulations. This type of system allows for greater flexibility without creating a mandatory cap for each site (Faeth 2000).

Woodward and Kaiser (2002) describe four major structural types of water quality trading markets including exchanges, bilateral negotiations, clearinghouses and sole-source offsets. Exchange markets are characterised by free exchange of information between buyers and sellers and a fluid transaction process. In this type of market there is a transparent market price at any point in time that both buyers and sellers can easily observe. Once an exchange market is initiated, the costs per trade are quite low. Start-up costs, however, can be quite high and thus there must be economies of scale for this type of structure to be efficient. A bilateral negotiation requires much more interaction between the buyer and seller before a trade can take place. Information is shared between the parties and negotiations take place before a trade agreement is reached. The transactions costs for this market structure are much higher than those for the exchange market but it allows for trading of goods that are not homogeneous in quality. Bilateral negotiations are the most common structure of water quality trading markets (Woodward et al. 2002).

In a clearinghouse structure, an intermediary purchases the pollution reduction credits and then can resell those credits to buyers who need to exceed their allowed pollution levels. This structure decreases transaction costs compared to the bilateral trading structure but it does require that legal authority be given to a government agency or other party to serve as the intermediary. Sole source offsets are not technically a trading programme in that they only involve one polluter. In this type of structure, a polluter is allowed to exceed its limit at one site if it reduces pollution by an equivalent amount at another site somewhere else in the watershed.

An added element of complexity to water pollution trading systems involves the type of pollution source that is being regulated. Trades can occur between point sources, between nonpoint sources or between point and nonpoint sources. Trading between the same type of pollution source (i.e. point/point trading or nonpoint/nonpoint trading) is generally simpler to deal with than trading between different types of sources. When trading between point and nonpoint sources it is generally recommended that a trading ratio be applied since nonpoint source reductions are considered to be more uncertain than those for point sources. Some have suggested that a trading ratio of greater than 2:1 be used (indicating that a reduction of two units of pollution from a nonpoint source is required to offset one unit of production for a point source) when trading between point and nonpoint sources (NWF 1999).

A concern that is often cited in the discussion of water pollution trading is the development of "hot spots" or certain areas within a watershed where water quality is highly degraded (NWF 1999). The concern here is that, while a trading system may improve water quality overall in a watershed, certain areas could experience declining water quality if the system is not set up properly. For example, if a polluter undertakes pollution reduction measures downstream of its facility in order to increase discharges, water quality downstream of the site may suffer.

3.2.5 Payments for Enhanced Water Quality

Ecosystems provide a variety of environmental services including hydrological benefits, reduced sedimentation, disaster prevention, biodiversity conservation, and carbon sequestration (Pagiola and Platais 2002). These important environmental services, which all help maintain water quality, are often unrecognised until they are compromised, for example, through rapid upstream development and deforestation. Many current regulatory programmes fail to provide the necessary (economic) incentives for landowners to undertake adequate conservation measures, which would help ensure environmental services for a larger set of beneficiaries.

In order to help preserve natural ecosystems and ensure the continued provision of environmental services, several countries have initiated programmes that provide payment for these services. In other cases, downstream user groups have spontaneously moved ahead directly to provide (monetary) incentives for upstream users to ensure water of sufficient quantity and quality downstream. Under these programmes typically private landowners who change their land management practices receive payment for the enhanced water (quality) services resulting from these changes in practices. These payments are typically sourced from the beneficiaries of the enhanced environmental services, usually the downstream water users. Examples for such programmes include forest environments and maintenance of watersheds in Costa Rica and direct payments to farmers for environmental conservation in the US. Additional pilot programmes are being undertaken or considered in Colombia, Nicaragua, the Dominican Republic, Ecuador, and El Salvador (Pagiola and Platais 2002).

Several key factors are important to consider when introducing a programme of payment for environmental services. Ensuring that payments are ongoing for the length of the conservation effort and targeted to those individuals that are undertaking the conservation activities is crucial. Moreover, it is important that the programme does not create perverse incentives, for example, a programme that pays for reforestation may encourage some landowners to cut timber on their land in order to receive payments for reforesting the area. It should also be made clear what services are provided by the conservation efforts. This type of transparency is more likely to gain support from potential buyers of the services. As with many other market-based systems, the appropriate institutions must be in place to collect the payments and ensure that the conservation measures are actually being taken.

3.3 Applications in The Americas

As discussed in the preceding section, the direct use of water rights to address water quality problems has been limited to date. While many of the market-based approaches to water quality do intrinsically include some form of water rights, they have generally not been directly used to deal with water quality issues. Two US programmes in Nevada serve as an example of how water rights can be used to deal with nonpoint source pollution. These voluntary water rights purchase programmes have allowed farmers to sell their water rights to government or nonprofit environmental groups to increase the amount of in-stream environmental flows while also potentially reducing the level of agricultural nonpoint pollution.

The Lahontan Valley water purchase programme allows the purchase of water rights by the Federal government from local farmers to provide water needed for the Lahontan Valley wetlands, which had begun to decline in acreage due to water demands from the Newlands irrigation project. The US Fish and Wildlife Service, the State of Nevada and the Nature Conservancy have purchased water rights with the intention of transferring water to the wetlands (Lovell et al. 2000). The second project, the Truckee River Water Quality Agreement, uses water rights along with pollution permits to deal with water quality problems in the area due to an expected increase in urban population in the next 15 years. The project allows for the purchase of water rights in order to enhance in-stream flows during critical periods in addition to the pollution discharge permit it issued to help deal with water quality problems in the area (Lovell et al. 2000). The Lahontan Valley project has been shown to be efficient in that the least productive rights have been sold to the government.

Market-based instruments, in particular pollution charges, have been used to deal with water pollution in the Americas with varying degrees of success. The United States has traditionally used command and control policies to deal with water pollution, although the role of market-based approaches, particularly effluent trading programmes, has been increasing. These programmes are often used along with current regulations to help improve the efficiency of water quality policies. While pollution trading programmes have been very successful for other media, success so far in water quality trading has been less notable, even in the United States where most of the programmes have been initiated (Faeth 2000). Although many programmes have initiated only a few trades so far, the potential for efficiency improvements seems to exist. Many state and local governments have recognised this potential as 11 additional demonstration programmes were initiated in 2003 alone. In the following, we discuss the application and results of some market-based programmes implemented to date.

Colombia initiated a water pollution fee programme in 1997 which charges polluters for Biochemical Oxygen Demand (BOD) and Total Suspended Solids (TSS). The programme established escalating charges for greater levels of effluent discharge. The revenues from the charges were distributed to the local environmental authority that used them to invest in additional environmental improvements such as municipal waste treatment plants. Organic waste was reduced by 36 percent and TSS were reduced by 52 percent in Eastern Antioquia's seven primary watersheds by the end of 1999 (Ambrus 2000). The programme's flexibility has been cited as one of the main forces behind its success. Twenty other regions in Colombia have begun to implement similar measures, although political disagreements have led to problems and delays in some areas.

In Sao Paulo, the largest industrialised area in Latin America, and third most populated city in the world, effluent charges, such as industrial sewage tariffs based on pollution content or load, have been in effect sporadically since 1983, resulting in marked reductions in BOD and suspended particulate emissions in a number of sectors (Shaman 1996).

In 1983 CETESB (Companhia de Technologia de Saneamento Ambiental) in Brazil introduced a tariff based on the pollutant contents of industrial sewage;

however, the programme could not be expanded due to a lack of fiscal resources to invest in treatment stations. While until 1990 the state negotiated with firms over treatment services in exchange for a service charge, since then the state has been allowed to recover the cost of monitoring, collecting and treating water discharge, based on a tax calculated on the average effluent by sector (Shaman 1996).

In 1989 the state Environmental Management Commission ascertained that the Tar-Pamlico River Basin in North Carolina was nutrient sensitive, due to low dissolved oxygen and algae blooms caused by excess nitrogen and phosphorous in the River. The excess nutrients were found to come primarily from nonpoint agricultural sources, with additional contributions made by wastewater treatment plants, industrial discharge and mining operations. A programme was initiated in the Tar-Pamlico River Basin that allows point-source polluters to trade with one another under a cap. If they are unable to keep their emissions below the cap, they can pay into a fund that supports a government sponsored nonpoint source reduction programme. The efficiency of polluters was found to increase in the first phase of the programme and discharge levels were met. A second phase will gradually reduce the allowable discharges (Faeth 2000).

One of the most successful water quality trading programmes in the US to date has been the Long Island Sound Nitrogen Credit Trading Programme in Connecticut. This programme allows point sources to trade with other point sources with the goal of reducing nitrogen levels and resolving the hypoxia problem in the Long Island Sound. The programme was passed by the state legislature in 2001 and began operations in 2002. Trading is allowed between 79 publicly owned treatment works, with the Connecticut Department of Environmental Protection (DEP) acting as a broker for the trading. This involves the treatment works operators selling and buying credits to the DEP, reducing transaction costs and allowing the DEP to have control over the market. Early estimates project cost savings over a command and control programme of $200 million (Kieser and Fang 2004).

The final example is a notable example because it employs trading between point and nonpoint sources. A programme was developed in 1997 in the Minnesota River Basin to allow trading of nitrogen and phosphorous between two point sources and nonpoint sources in the watershed. The point sources have set up a trust fund, which provides funding for the programme to ensure that wastewater discharges are offset by reductions in nonpoint source pollution. The trading scheme utilises a trading ratio of at least 2:1 to take into account uncertainty of the nonpoint pollution control measures (Fang and Easter 2003). The Minnesota Pollution Control Agency closely monitors the programme to assure accountability. There have been five major trades and many smaller trades since the programme began. Estimates have shown this programme to increase cost efficiency, although the results vary depending on the nonpoint pollution control method used (Kieser and Fang 2004).

Payments for environmental services have been used in several areas in Latin America. One of the most oft-cited programmes is the Costa Rican Payments for Environmental Services Programme (PESP). The programme attempts to encourage forest conservation on private land by providing payment for watershed protection, carbon sequestration and biodiversity, as well as for tourism and scenic beauty. Financing for this programme comes primarily from a tax on fossil fuels, with additional funding from payments from private sector firms for watershed conservation, the sale of carbon offsets and international donations from the Global Environment Facility.

Other areas have experimented with programmes that receive funding from municipal government or water user groups. In Quito and Cuenca, Ecuador, municipal water authorities are providing funding for conservation activities in the upper part of the watershed from which they receive most of their water. In Valle del Cuaca, Colombia, a group of rice farmers formed the Guabas River Water User Association in order to address water scarcity in their region. Initially, the association bought land upstream that was threatened by erosion. A more recent programme pays upstream landowners for watershed management activities using fees paid by association members (Landell-Mills 2002).

3.4 Constraints and Solutions

As seen from the above examples and discussion, although market-based solutions to water quality have the potential to reduce pollution and costs, many constraints still exist. Some of the major concerns to be considered include institutional constraints, enforcement problems, and economies of scale and "hot spots" in the case of effluent or nutrient trading markets. Although many of these constraints apply to market-based instruments in any country, we will also focus on the specific constraints developing countries face.

There are many institutional constraints faced in developing countries that should be taken into account when considering market-based approaches to water quality. While some of these constraints are specifically associated with market-based approaches, others would apply to command and control environmental regulations as well. Limitations that may be particularly prevalent in developing countries, such as lack of funding, inexperience, unclear jurisdiction and lack of political will, all influence the ability of the programmes to succeed.

One of the first hurdles to overcome when implementing market-based instruments for water quality protection is lack of political support. This is often considered to be a problem in developing countries, but it has been a problem in developed countries as well. In the case of water pollution trading markets, farmers have resisted the inclusion of nonpoint source pollution. Environmental groups have also opposed them on ethical grounds.

Grandfathering is one way to garner support from firms already in the market when initiating a new pollution control programme. Another strategy that has generally led to greater support is the gradual implementation of effluent fees. Initially, the programme can charge only non-compliance fees, and then later assess a fee for discharges within the set standards (O'Connor 1999). Providing information to the public about the programme can also help to increase support for new pollution fees. Earmarking fees for pollution control measures can also help to in-

crease public support, although, as mentioned earlier, care should be taken to ensure the tax is not set too high.

Although the implementation of pollution fees can provide some level of funding as described above, lack of funding for environmental protection is a fundamental problem in many countries. Market-based instruments are anticipated to
reduce costs. However, this may not always happen in practice. Faeth (2000) notes
that administrative and transaction costs have actually increased in some programmes, making trading difficult.

Lack of institutional capacity is a primary concern in most developing countries
when instituting market-based or command and control regulations. However, it
has been noted that many developed countries may experience some of the same
institutional constraints (Serôa da Motta et al. 1999). Many reforms that are currently taking place in developing countries, including the removal of price distortions, increased trade liberalisation, and other market reforms, will likely increase
the chance of success for many market-based instruments (O'Connor 1999).

Monitoring can be a problem for market-based and command and control regulations for both developed and developing countries. In developing countries baseline environmental data is often lacking, making it difficult to document any environmental improvements. In addition, many countries lack the capacity to monitor
pollution sources. Nonpoint source monitoring is particularly difficult as the
amount of pollutants reduced is generally an estimate. The actual amount of pollution can vary, based on a number of factors, including weather, various location-
specific characteristics and how well the management practice is implemented
(Woodward et al. 2002). Monitoring costs can also be quite high, which can affect
developed countries but could have a profound impact on the ability of developing
countries to implement water quality improvement policies. Monitoring costs are
required not only for traditional command and control measures for water pollution control but for market-based instruments as well.

One of the major issues that must be addressed with any kind of water quality
regulation is enforcement. Many times regulations are in place but pollution reduction goals are not reached due to a lack of enforcement. Inadequate enforcement of environmental regulations is a hindrance in many developing countries.
Although some authors suggest that increased use of market-based instruments
might remedy this problem, O'Connor (1999) points out that it depends on the underlying cause of the enforcement difficulties. If enforcement is weak because of a
lack of political commitment to water quality goals, then the same enforcement
problems are likely to exist with market-based instruments or with command and
control approaches. Incentives could help increase enforcement, however, if the
problem is that government agents do not have the proper incentives for strict enforcement under the current regulations.

Although water quality trading programmes have been initiated in several watersheds around the US, relatively few trades have occurred to date. As of 1999,
only 10 trading events had occurred since the inception of water quality trading
programmes in the United States, with only one or two trades occurring in each of
the programmes that had been initiated more than a decade before (Woodward
2003). More recent data suggests more trades have taken place in some newer

programmes but the overall level of trading is still limited (Kieser and Fang 2004). This lack of trading could negate the possible cost saving benefits of these programmes. One reason that so few trades have been occurring is due to the nature of water pollution. Water pollution affects a smaller area than air pollution; therefore water pollution trading programmes must be set up within watershed boundaries. This limits the number of possible participants in the trading market, making it difficult for polluters to identify acceptable trades (Woodward 2003).

Water pollution effects can be variable depending on where and when they are emitted. A pollutant's effect on a watershed can be influenced by weather, soil conditions, slope, land use and other physical characteristics. This leads to a concern about "hot spots" or localised areas of pollution that could occur. In some situations, trading could lead to a decrease in overall pollution in a watershed, while certain areas within the watershed could experience increased pollution if emissions increased in a given area. Due to the concern about hot spots, trading of toxic pollutants is not considered appropriate. Precautions can be taken in the planning stages of a trading programme to guard against hot spots. The location of trading partners within the watershed, the size of the trading area, trading ratios between point and nonpoint sources, enforcement and monitoring should all be carefully considered when designing a trading programme (NWF 1999).

3.5 Road Map for the Future

Water quality improvements in the Americas face many constraints. But these improvements are crucial to ensuring public health and the availability of adequate supplies of clean water. The main steps forward for improving water quality in the Americas include increased investments for enhanced sanitation and the implementation of water quality policies, education and awareness-building for both rural and urban users, and increasing the institutional capacity to deal with water quality problems. The market-based tools discussed in this chapter can be useful in meeting water quality goals but should be carefully considered in the context of each country.

Strengthening institutional capacity is one of the first factors that should be addressed to help ensure the implementation of water quality policies. Government agency roles should be clearer to organise the implementation, monitoring and enforcement of environmental policies better. Lack of experience and funding also impact the ability of governments to address water quality issues. While general lessons can be learned from experiences in developed countries, it is crucial that the water pollution policies for developing countries address their specific challenges and do not simply apply OECD country approaches. Historically, much of the information about environmental regulations and market-based instruments has come from countries in the north. Information sharing between countries in Latin America and the Caribbean is crucial to enhance the success of water quality improvements.

Public education and awareness-building about water quality issues is essential to helping gain public acceptance of water quality policies and government investments in this area. This can help increase political support of water quality policies. Education efforts can also help reduce public health impacts of water quality problems by increasing awareness of measures that can be taken to decrease health risks.

Large increases in investments are needed to implement effective water pollution control policies and adequate monitoring and enforcement programmes to support them. Additional funding will also be needed for improvements in institutional capacity and educational programmes. According to baseline estimates from the International Model for Policy Analysis of Agricultural Commodities and Trade (IMAPCT), $9.8 billion would be required between 1997 and 2020 to improve access to clean water in Latin America and the Caribbean to 81 percent, up from 77 percent in 1997 (Rosegrant et al. 2001). While pollution charges could provide some of this funding, additional funding is likely to be required. Donor support and private investments in water supply and sanitation may also be needed to obtain the level of investment required to reach water quality goals.

The market-based instruments discussed in this chapter can be useful for reaching water quality goals but should be geared to the economic and political conditions in each country, and be introduced gradually with the flexibility necessary to adjust to institutional changes occurring in the region. Combining command and control regulations with market-based methods will likely provide the best results in terms of achieving water quality and economic efficiency goals.

3.6 References

Ambrus S (2000) Colombia Tries a New Way to Fight Water Pollution…and it Works. EcoAmericas. World Bank, Monthly Report (March)

Boyd J (2003) Water Pollution Taxes: A Good Idea Doomed to Failure? Resources for the Future Discussion Paper 03-20, Washington, DC

Davis R, Hirji R (2003) Water Quality: Assessment and Protection. Water Resources and Environment Technical Note D.1, World Bank, Washington, DC

Faeth P (2000) Fertile Ground: Nutrient Trading's Potential to Cost Effectively Improve Water Quality. World Resources Institute, Washington, DC

Fang F, Easter KW (2003) Pollution Trading to Offset New Pollutant Loadings: A Case Study in the Minnesota River Basin. (Paper presented at the American Agricultural Economics Association Annual Meeting, Montreal, Canada)

Guigale MM, Lafourcade O, Luff C (2003) Colombia: The Economic Foundation of Peace. World Bank, Washington, DC

Huber RM, Ruitenbeek J, Serôa da Motta R (1998) Market Based Instruments for Environmental Policymaking in Latin America and the Caribbean: Lessons from Eleven Countries. Discussion Paper no. 381, World Bank, Washington, DC

Kieser MS, Fang F (2004) Economic and Environmental Benefits of Water Quality Trading: An Overview of U.S. Trading Programmes. (Paper presented at Workshop on Urban Renaissance and Watershed Management, Tokyo and Otsu, Japan)

Landell-Mills N (2002) Marketing Forest Environmental Services – Who Benefits? International Institute for Environment and Development Gatekeeper Series, No. 104, London

Lovell S, Millock K, Sundig DL (2000) Using Water Markets to Improve Environmental Quality: Two Innovative Programmes in Nevada. Journal of Soil and Water Conservation 55(1): 19-26

NWF (1999) A New Tool for Water Quality: Making Watershed-Based Trading Work for You, National Wildlife Federation, Washington, DC

O'Connor D (1999) Applying Economic Instruments in Developing Countries: From Theory to Implementation. Environment and Development Economics 4(1): 91-110

Pagiola S, Platais G (2002) Payments for Environmental Services. Environmental Strategy Notes, No. 3, World Bank, Washington, DC

Rosegrant MW, Paisner MS, Meijer S, Witcover J (2001) Global Food Projections to 2020: Emerging Trends and Alternative Futures. International Food Policy Research Institute, Washington, DC

Serôa da Motta R, Huber RM, Ruitenbeek HJ (1999) Market Based Instruments for Environmental Policymaking in Latin America and the Caribbean: Lessons from Eleven Countries. Environment and Development Economics 4(2): 177-201

Shaman D (1996) Brazil's Pollution Regulatory Structure and Background. http://www.worldbank.org/nipr/brazil/braz-over.htm

Stavins RN (2000) Market-based Environmental Policies. In: Portney PR, Stavins RN (eds) Public Policies for Environmental Protection. RFF Press (Resources for the Future), Washington, DC

WHO and UNICEF (World Health Organisation and United Nations Children's Fund). (2000) Global Water Supply and Sanitation Assessment 2000 Report, Geneva, Switzerland

Woodward RT (2003) Lessons about Effluent Trading from a Single Trade. Review of Agricultural Economics 25(1): 235-245

Woodward RT, Kaiser RA (2002) Market Structures for U.S. Water Quality Trading. Review of Agricultural Economics 24(2): 366-383

Woodward RT, Kaiser RA, Wicks AB (2002) The Structure and Practice of Water Quality Trading Markets. Journal of the American Water Resources Association 38(4): 967-978

Effectiveness of Market Approaches to Water Quality Management

K. William Easter and Robert C. Johansson[1]

4.1 Introduction

In his article promoting the access to safe drinking water as a human right, Gleick (1999) cites that as many as 30,000 people per day perish from water-related illness. Moreover, global stressors on human access to basic water provision are expected to increase in the 21st century, influencing both quantity and quality aspects of water security. Two thirds of people are expected to be living under water-stressed conditions by 2025 (UNEP 1999); municipal and industrial uses of water have grown by 24 times since the last century and urban populations are expected to swell to 5 billion by 2025 (Gardiner 2002); and global food production is expected to become increasingly intensive to meet increasing demand. For example, a major source of water quality degradation is the discharge of organic chemical compounds from production activities, which result in biochemical oxygen-demand, or BOD. In 23 selected countries in the Caribbean, Latin and North America, these pollutants summed to more than 4 million kilograms of biochemical oxygen demand (BOD) per day, an average of about 55 kg of BOD per worker per year (see Table 4.1). The majority of this discharge comes from the production of food and beverages (44 percent). In the United States an estimated 750,000 kg of BOD per day are generated in the production of food and beverages. BOD or BOD-contributing substances cause a majority of the 29,711 impaired water bodies in the United States. These waters encompass 300,000 miles of rivers and shorelines and 5 million acres of lakes, which are located within 10 miles of 218 million people (EPA 2005).

In part, water quality degradation results from unclear or nonexistent property rights for clean water; i.e., polluters do not account for the harmful effects of their pollutant discharge (e.g., organic chemical compounds, fertilisers, pesticides, heavy metals, sediment) when making production decisions. When polluters do incorporate the harmful effects of their pollution into production decisions, the net result is not necessarily a total cessation of pollution. This is because positive quantities of pollution may occur in a socially-efficient outcome, one that maximises the difference between total benefits resulting from cleaner waters and the total cost of achieving reductions in discharge (Coase 1960; Arrow and Hahn 1971).

---

[1] The views herein are those of the authors and not necessarily those of the Economic Research Service or the United States Department of Agriculture.

Table 4.1. Discharge of organic chemical compounds from production activities (World Bank 2003). Data refer to any year from 1993 to 2000; data for the Dominican Republic and Nicaragua was taken from 1980. Industry shares may not sum to 100 percent because data may be from different years.

	Emissions (kg)		Industry Share (%)							
	day^{-1}	day^{-1} worker^{-1}	Primary metals	Paper & pulp	Chem.	Food & Bev	Ceramics & Glass	Textiles	Wood	Other
Argentina	177,882	0.21	7.06	11.62	8.00	58.97	0.23	8.40	1.80	3.88
Bolivia	12,759	0.25	0.85	20.46	7.00	61.39	0.26	7.14	2.39	0.90
Brazil	629,406	0.20	10.51	14.06	9.00	42.67	0.35	14.47	3.47	6.87
Canada	307,325	0.15	10.76	23.93	10.00	34.84	0.13	5.42	5.09	10.04
Chile	72,850	0.24	6.85	11.32	9.00	62.72	0.12	4.99	2.58	2.49
Colombia	100,752	0.21	3.93	16.23	10.00	51.12	0.19	14.81	0.71	2.73
Costa Rica	35,164	0.22	1.42	9.52	7.00	64.28	0.12	13.36	1.60	2.57
Dominican Republic	54,935	0.38	0.64	2.78	1.89	92.14	0.08	1.94	0.23	0.31
Ecuador	32,266	0.27	2.05	10.85	6.00	65.48	0.20	9.65	2.20	2.46
El Salvador	22,760	0.18	3.45	13.24	8.00	57.94	0.11	16.44	0.47	1.24
Guatemala	19,253	0.28	2.35	10.12	6.00	72.81	0.23	9.76	1.27	1.01
Honduras	34,036	0.20	1.10	7.83	4.00	55.48	0.14	26.80	4.04	0.77
Jamaica	17,507	0.29	6.93	7.24	4.00	70.84	0.11	9.76	1.34	0.01
Mexico	296,093	0.20	7.79	12.52	10.00	55.56	0.19	7.47	0.92	5.13
Nicaragua	9,647	0.28	0.23	5.43	5.08	79.68	0.09	7.60	0.98	0.91
Panama	11,461	0.31	2.21	14.08	5.00	67.96	0.20	8.97	1.83	0.94
Paraguay	3,250	0.28	2.28	9.93	6.00	73.61	0.27	6.73	0.33	0.86
Peru	52,644	0.21	7.94	14.01	9.00	47.24	0.23	14.10	2.17	3.76
Puerto Rico	16,207	0.14	0.99	13.53	19.00	37.57	0.18	17.29	1.57	9.38
Trinidad & Tobago	11,787	0.28	4.43	14.60	7.00	51.59	0.26	8.79	2.18	1.15
USA	1,968,196	0.12	10.50	10.97	14.00	38.41	0.21	7.11	4.10	14.93
Uruguay	23,109	0.27	3.44	11.23	6.00	72.26	0.19	6.59	0.73	1.76
Venezuela	94,175	0.21	13.66	13.91	10.00	46.91	0.24	9.86	1.71	3.89
Average		0.15	9.33	12.73	11.38	44.33	0.22	8.83	3.35	10.12

Policy makers use several mechanisms to induce polluters to incorporate the social costs of their pollution into their profit-maximizing production decisions. Often times due to incomplete knowledge of the benefits of cleaner water and the costs associated with pollution abatement, the policy maker will simply choose a pollution standard and mandate that polluters uniformly reduce their discharge equal to that level. Market-based alternatives to this command-and-control approach are thought to be more cost-effective and have been suggested to policy makers by academics for many years. The use of markets and associated mechanisms to reduce pollution offer the possibility of managing water quality in a cost-effective manner. If the rights to clean water (or alternatively to pollute) are properly defined and when these rights are tradable (as in the case of tradable pollution permits) a market equilibrium between buyers and sellers of pollution permits can achieve the desired water quality goal at least cost.

Stavins (2003) loosely categorises market-based approaches to improving environmental quality into four categories: charge systems (a price instrument), tradable permits (a quantity instrument), market friction reduction (essentially addressing informational needs), and government subsidy reduction (another price instrument). In this chapter, we focus our discussion on permit markets for water quality between point source (PS) dischargers (e.g., wastewater treatment facilities) and nonpoint source (NPS) dischargers (e.g., cropland). However, because obstacles often exist to "properly defining" the rights to pollute or to governing their sales and purchases, we also discuss these market frictions, or transaction costs, associated with tradable permits for water quality.

In the first section, we briefly touch on the literature behind permit trading and provide a simple numerical illustration of the equi-marginal cost principle underlying this approach to environmental management. Next, we discuss the literature on obstacles to permit trading and how these may be addressed. In this context, we consider several current permit-trading policies for water quality. We conclude with discussion of conditions that might facilitate the evolution of market-based systems.

4.2 Background

The theory behind permit markets stems from the seminal work of Dales (1968) and Montgomery (1972), from which many types of marketable permit systems have arisen (Baumol and Oates 1988; Hanley et al. 1997). The generic cap-and-trade permit system for air pollutants is commonly referred to in the literature as an emissions trading system, which we term a discharge trading system (DTS) for water quality management. Under this system, aggregate discharge of the target pollutant is constrained to an environmental standard, which is set by the regulatory agency. Permits, representing the right to discharge a certain quantity of that pollutant, are distributed to the sources of the pollution, such that the total number of permits sum to the environmental standard. Sources buy and sell these permits on the open market; so long as their discharge of the pollutant in some time period are less than, or equal to, the number of permits held, they are not subject to any penalties.

Other types of permit systems include ambient permit systems (APS), pollution-offset systems, non-degradation offset systems, and modified pollution offset systems. An APS weights the expected environmental impact of discharge from different sources and is appropriate when the amount of discharge and its spatial distribution are important to consider (Horan et al. 1998). Offset systems are hybrids of the DTS and APS allowing for different degrees of control over the timing, spacing, and quantities of discharge (Hanley et al. 1997). Under these systems the marginal cost of abatement will be equalised across sources such that the equilibrium price of purchasing an additional permit will equal the marginal cost of abating an additional unit of pollution. Those sources having marginal abatement costs greater than the permit price will prefer to abate less and to purchase additional permits and vice versa.

As an example, consider the case of the Sand Creek, a tributary of the Minnesota River (Figure 4.1), where excessive amounts of phosphorus (P) are discharged into the River from surrounding cropland, livestock facilities, and water treatment plants (Johansson et al. 2004).

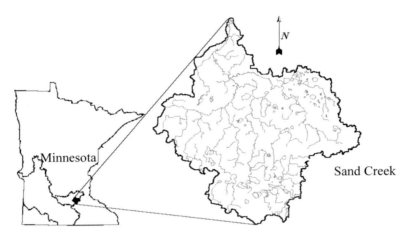

Fig. 4.1. Sand Creek watershed in the State of Minnesota (Johansson et al. 2004)

Here point sources (m) currently discharge 40,000 lbs of P per year (e_m) into the Minnesota River. The cost of abating point source discharge, $C(a_m)$, is $0.005(a_m)^2$ where a_m=40,000-e_m. There are also 150,000 acres of cropland (n) that discharge 0.5 lbs of P per acre per year into the River (e_n). The cost per acre of abating nonpoint discharge, $C(a_n)$, is $134.35(a_n)^2$ where a_n=0.5-e_n. The policy maker wishes to reduce total discharge into the River by 40 percent, a level not achievable by point sources alone.

Under a uniform reduction policy, each source reduces discharge by 40 percent: the point sources abate 16,000 lbs of P per year and cropland abates 30,000 lbs of P per year. Point source abatement costs are $0.005(16,000)^2$, or $1.3 million. Nonpoint source abatement costs are $134.35(0.2)^2$, or $5.37 per acre. Total costs are $1.3 million + 150,000($5.37), or $2.1 million. The average cost per pound of P-abatement (excluding potential transaction costs) equals $46.

Under a discharge trading system, the regulator distributes 69,000 permits to the point sources and to the nonpoint sources, where each permit (ℓ) represents the right to discharge 1 lbs of P into the River that year. The regulator distributes these permits according to historic levels of pollution such that each polluter receives permits equivalent to 60 percent of historic discharge levels: ℓ_m = 24,000 lbs and ℓ_n = 0.3 lbs per acre. Denote the purchase or sale of a permit by S. Each source will choose abatement and sales to minimize $C(a_i)+P_\ell S_i$ subject to $a_i \geq 0$, S_i =e_i-ℓ_i-a_i, i=m and n, and P_ℓ is the market clearing price for permits.

The market-clearing conditions are: $a_m=100P_\ell$, $a_n=0.004P_\ell$, and $a_m+a_n\geq46,000$. Solving, the price for permits will equal $70 with $a_m=7,000$ lbs and $a_n=0.26$ lbs per acre. The resulting annual abatement costs would be $0.005(7,000)^2$, or $245,000 for point sources and $134.35(0.26)^2$ per acre, or $9.08 per acre for nonpoint sources. Total costs would be $1.6 million with an average cost per pound of P-abatement (excluding potential transaction costs) of $35. Nonpoint sources would sell 9,000 permits to point sources at a price of $70; the net cost to the point sources is $875,000 and is $4.88 per acre for cropland.

Comparing the two policies in this example, the average cost per pound of P-abatement is a 24 percent less under a trading system than under a uniform reductions policy. Similar abatement could be achieved with an effluent fee of $70 per pound of phosphorus discharged, but such a policy would likely be unacceptable to point and nonpoint sources. Under a policy using effluent fees, the wastewater facility would pay $490,000 in fees and would incur $245,000 in abatement costs, a total of $735,000. Cropland would pay on average $16.80 per acre in fees and would incur an additional $9.08 per acre in abatement costs. It is not surprising that farmers (and other polluters) are opposed to effluent fees, even though the tax revenues may be used to offset abatement costs (McCann 1997).

4.3 Transaction Costs and Uncertainty

The tradable permit process will ensure the lowest cost of compliance in achieving the predetermined level of abatement excluding transaction costs. Transaction costs include such as those required for monitoring of discharge and enforcement of the environmental standard (Kaplan et al. 2003). Stavins (1995) has shown that these transaction costs are not negligible for permit markets. However, he concludes that even if transaction costs prevent a permit system from realizing a high number of trades, the aggregate costs of compliance will likely be less expensive than a uniform reductions approach commonly used in command-and-control regulation. Furthermore, because permit trading encourages innovation across all sources and technologies (i.e., so that a source can sell more permits or purchase fewer), a permit system where no trades occur is also likely to be less costly than a fixed technology standard (O'Neil et al. 1983).

Many authors have built on the deterministic permit model to reflect real world complications and to facilitate policy comparisons. Under such "second-best" conditions that include transaction costs, mechanisms such as standards may result in the lowest overall cost of regulation (Stavins 1995; Smith and Tomasi 1995; McCann and Easter 1999; Johansson 2002). Consider a broad definition of transaction costs, one defined as the resources required to define, establish, maintain, and transfer the property rights (Table 4.2). In the case of water pollution, it would involve the cost of establishing the right to discharge a pollutant, such as a pound of nitrogen or phosphorus, as well as the cost of transferring the rights from one owner to another. In some cases, not all the costs will be relevant, while in others, all will be important. The importance of specific transaction costs will vary by type of pollutant, its source and existing institutional arrangements.

Table 4.2. List of transaction costs

Transaction Cost	Description
Research and information	Defining the water quality problem and sources
Enactment or litigation	Soliciting public acceptance of the problem and solution(s)
Design and implementation	Soliciting polluter acceptance of the problem and solution(s)
Support and administration	Setting up the institutional framework to impose the solution(s)
Contracting	Polluter responses to the solution
Monitoring and detection	Information gathering to re-examine the problem and solution(s)
Prosecution and enforcement	Imposing solution on unwilling sources

These transaction costs are essentially the costs incurred by the policy maker or the polluter to address one form of uncertainty or another (Taff and Senjem 1996): uncertainty about spatial variability of water quality (physical uncertainty), uncertainty about source behaviour (behavioural uncertainty), uncertainty about the weather (stochastic uncertainty), or uncertainty over time (dynamic uncertainty). For example, uncertainty about polluter behaviour under regulation gives rise to monitoring, detection, prosecution, and enforcement transaction costs. Similarly, research, information, design, implementation, and monitoring transaction costs are a function of how much uncertainty there is surrounding the water quality problem and how it may be affected by abatement efforts and weather.

4.3.1 Physical Uncertainty

Knowledge of the variable impact of abatement efforts at different geographic locations is necessary to implement such a policy cost-effectively. This speaks initially to research and informational costs, but also later to monitoring costs. For example, are reductions of a pollutant, such as sediment, in one region more or less valuable than reductions elsewhere? In the point-nonpoint offset between the Rahr Malting Company and agricultural NPSs (MPCA 1997) for the Minnesota River, trades were defined for phosphorus and nitrogen in terms of the expected secondary impact on the River's water quality. Nutrients were converted to a normalised unit (pound equivalents of $CBOD_5$[2] removed from the water). It was then estimated that 1 lb of phosphorus prevented from entering the Minnesota River near its confluence with the Mississippi River resulted in 8 lb of $CBOD_5$ abatement. However, phosphorus was more limiting for algae growth upstream, where

---

[2] Five-day carbonaceous biochemical oxygen demand.

1 lb of phosphorus abatement was estimated to generate as much as 17 lb of $CBOD_5$ abatement. When trades are allowed across larger areas, the information costs associated with a larger geographic area will be of increasing importance (Kaplan et al. 2003).

4.3.2 Behavioural Uncertainty

Uncertainties about how sources will behave under various market structures lead to other transaction costs: initially, enactment, design, and support; later, detection, enforcement, and prosecution. Taff and Senjem (1996) discuss uncertain participation in water quality markets mainly as a function of source uncertainty. For example, do buyers that wish to purchase permits solicit individual sellers (increasing individual contract costs), or do buyers and sellers connect on an online auction, such as the World Resource Institute (WRI 2004) pilot programme in Kalamazoo, MI (increasing policy makers' design and implementation costs?) If the participating farmers do not achieve the expected level of reduction due to a large rainfall event (stochastic uncertainty, discussed below), will they be penalised (increasing policy makers' monitoring and enforcement costs and increasing individual research and information costs)? Once all of these issues have been clearly laid out, the various sources can weigh the costs and benefits of whether to participate or not. The larger the required reductions in discharges and the greater the differences in abatement costs among sources, the higher will be the opportunity costs of non-participation.

Under imperfect information, adverse selection and moral hazards may be present, such that sources may have incentives to misrepresent abatement efforts (Xepapadeas 1991). Research has illustrated how mechanism design might mitigate these incentives by constructing a complex tax structure that includes fees for programme implementation (Smith and Tomasi 1995). An alternative to mechanism design is a conventional market structure coupled with a rigorous monitoring programme, which would reveal non-compliance (Kaplan et al. 2003). However, the costs of such monitoring may outweigh the subsequent gains (Johansson 2002).

Another method with which to induce polluters subject to moral hazard to accept full responsibility for their marginal contribution to environmental damage is to penalize, or reward, them on the basis of ambient environmental quality, which is relatively easy to measure. Segerson (1988) discusses the problem of assigning responsibility to NPS when only observations of the ambient environmental quality are known. A range of expected discharge can be modelled using a probability density function, which is conditioned on the adopted abatement practice. This approach is generalised in Horan et al. (1998), when farmers have multiple abatement strategies available to them. However, while ambient water quality is relatively easy to measure, setting penalties and awards for individual actions based on a group's actions involves relatively high transaction costs. It is extremely difficult for individuals to ascertain the behaviour of other group members as they make their production and abatement choices.

Similar to this notion of using ambient water quality to alter programme design, is the notion of employing "adaptive management," which allows an iterative process to adjust policy-given observations of past policy performance. Moledina et al. (2003) illustrate how the regulator can employ such a strategy in a repeated game framework to mitigate moral hazard in regulated players. The ability to re-formulate market structures in this way also provides a means to prevent market power distortions in permit markets (Hahn 1984; Hagem and Westskog 1998).

4.3.3 Stochastic Uncertainty

This refers to the policy maker's uncertainty about how stochastic events, such as weather and the economy may influence the outcome of a particular policy. For example, in a drought year (without irrigation) a typical farm may discharge just as much fertiliser or pesticide into surrounding surface water under a market-based abatement policy as under one of uniform reductions command-and-control. Compounding the impacts of stochastic weather on abatement efforts are the sometimes adverse impacts on agricultural production. In an extremely wet year, farm discharge could be higher, but agricultural production lower, than expected.

Such uncertainty serves to increase the cost and importance of design, monitoring, and implementation (in the case of adaptive management). Weitzman (1974) considers stochastic uncertainty and choice of optimal price or quantity instrument in his seminal article, "Prices –vs- Quantities." In Wietzman's model the planner can choose a price-instrument (e.g., effluent fee) that induces the producer (polluter) to supply an optimal level of the economic variable (pollution abatement), or the planner can use a quantity instrument to restrict the production of a pollutant at an optimal level. Weitzman shows it makes no difference in terms of social welfare whether the planner uses price or quantity instruments in an environment of perfect information[3]. However, when the planner is uncertain about the benefits or costs as a function of the production of the economic variable, conditions exist in which the planner should choose one instrument over the other.

Typically, stochastic influences on discharge trading systems have been addressed by basing trades on expected discharge and weighing source discharges by their degree of certainty. In the case of PS discharge, such as effluence from wastewater treatment facilities, the degree of uncertainty around discharge is relatively low; in the case of NPS discharge, there is a relatively high degree of uncertainty. Furthermore, because NPS discharge is difficult to measure and assign ownership, expected discharge is often estimated using physical process models. In cases where sources with different uncertainties are trading among each other, many have examined the degree to which discharge from one source should be equated to others using "trading ratios" (Malik et al. 1993).

---

[3] His analysis does not include transaction costs or the cost to farmers of paying effluent fees in addition to incurring abatement costs. Weitzman points out that these issues are concerned with the implementation of policies and not with the comparisons of the policies themselves.

4.3.4 Dynamic Uncertainty

When a pollutant does not flush from a water body or degrade in a timely fashion, it is considered a stock pollutant, because discharge can build up over time. In the case of stock pollutants, it is important to consider multi-period policies and adaptive management techniques. Hoel and Karp (2001) have extended the Weitzman treatment of uncertainty to a dynamic environment in which pricing and quantity instruments can be compared over time. Hoel and Karp allow this uncertainty to enter both additively and multiplicatively. The cost and benefits functions not only shift up and down, i.e. additively, as in Weitzman (1974), but also may have slope changes (i.e. multiplicatively). The conclusions drawn in this analysis are quite similar to Weitzman; whether taxes dominate quotas will primarily depend on the slopes of the cost and benefits functions.

4.4 Applications

The potential to generate significant nutrient abatement at low cost using tradable permits has led to several experimental programmes encouraging NPS offsets in lieu of increased PS controls – at least 37 prior to 1999 (Environomics 1999). These programmes, created in response to underlying water quality problems, developed subject to the relevant transaction costs discussed above and fall into one of three market types (Woodward and Kaiser 2002): exchanges, bilateral negotiations, and clearinghouses. Exchanges (essentially reflecting the discharge trading system example above) represent the purest form of a market and envision trades for effluent offsets occurring on the open market similar to SO_2 offsets under Title 4 of the Clean Air Act. In order for exchanges to operate effectively, the effluent abatement underlying a permit would have to be uniform and verifiable across space. Bilateral negotiations are market transactions between an individual permit buyer and seller. Woodward and Kaiser list several examples of these types of trades that develop primarily in response to stochastic uncertainty and asymmetric information; i.e., the underlying permit base is difficult to estimate, monitor, and enforce. These include the Wisconsin Fox River Programme, California's Grassland Drainage Area Programme, and the WRI experimental online nutrient trading programme (WRI, 2004). Water-quality clearinghouses are another type of market structure, whereby intermediary brokers permit trades among a variety of buyers and sellers. An example of this discussed below is the Tar-Pamlico trading programme.

However, before turning to a closer examination of individual programmes, it bears mentioning what researchers have concluded about the successes or failures of past market programmes (EPA 2002c) and their future potential. King and Kuch (2003) pose the question, "Will Nutrient Credit Trading Ever Work?" In their examination of 37 markets, designed to address excess nutrient discharge to surface water, King and Kuch distil their answer from a comparison of nutrient credit supply and nutrient credit demand. They conclude that to employ permit trading successfully for water quality management, the number of potential buyers and sellers of pollution abatement must exceed a certain critical mass, which has yet to be achieved under water quality trading markets. These participants must have heterogeneous costs of pollution abatement. The system must be designed to allow a large number of heterogeneous participants. All of these requirements can be met, argue King and Kuch, by water quality managers, who can pass along the cost of setting up the institutional mechanisms to taxpayers in general. They note two conditions that restrict the general use of markets for water quality in the United States: farmers are already receiving payments for restricting nutrient discharge via conservation programmes and point source discharge has not yet been restricted sufficiently to generate widespread demand for a nonpoint source market for offsets. A recent analysis of one of the oldest markets for water quality management in the United States, Lake Dillon Reservoir in Colorado, confirms many of these conclusions (Woodward 2003). After being designed in 1984, the Lake Dillon programme witnessed no actual trades until 1999. Woodward cites as a reason lack of demand due to technological innovation and inflexible market structure (essentially supply restrictions).

In focusing on supply and demand-side considerations of water quality markets, King, Kuch, and Woodward have helped identify necessary conditions for markets, but these are often not sufficient. In addition to the necessary supply and demand conditions, it is also necessary that the overall efficiency gains to trading outweigh the implicit transaction costs incurred due to uncertainty, when compared to alternative water quality management options. One must ask the question, "Will trading increase water quality more or less than another form of regulation?" and "Will trading result in higher or lower abatement costs – both for polluters and for policy implementation?" With PS and NPS abatement policies there will be uncertainty regarding the ability of management practices to meet TMDL allocations and subsequent costs. The policy maker must balance these uncertainties in selecting the appropriate policy for a given watershed, based on the number and type of sources and the degree to which abatement costs differ between sources.

4.4.1 Total Maximum Daily Loads (TMDLs) – A Framework for Markets?

The Clean Water Act essentially requires individual state agencies to monitor and list impaired rivers, lakes, and estuaries. States are also required to develop (with the assistance of the EPA) a water quality management plan, or TMDL, to address the designated impairments. These have been developed for a number of the 29,711 impaired waterways within the United States (EPA 2005). Many TMDLs rely on command-and-control effluent standards or technologies for point sources through the National Pollutant Discharge Elimination System, authorised under the Clean Water Act. However, of those impaired waters surveyed, 43 percent are attributable solely to diffuse, nonpoint sources (NPSs) such as agricultural lands, and 47 percent have both NPS and point source (PS) contributions (EPA 2002a). This indicates that NPSs will need to be an important component in United States efforts to create markets for water quality with respect to eutrophication (excess nitrogen and phosphorus discharge), hypoxia (low dissolved oxygen due to excess nutrient loads), sedimentation and pesticides. The key problem is to develop property rights for the pollutants that can be traded between point and nonpoint sources (research and information costs). Furthermore, in the United States only point sources have been required to reduce pollution discharges (enactment or litigation costs).

To date, relatively few of these water quality management plans have been developed and implemented, for many of the reasons mentioned above: high transaction costs due to uncertainty and thus far no binding effluent restrictions for NPSs. To facilitate TMDL implementation, EPA has shown a willingness to accept some of the transaction costs inherent in permit trading programmes (EPA 2002b), including:

- Improve state monitoring and assessment programmes that support the TMDL programme.
- Strengthen watershed planning processes to foster TMDL implementation.
- Increase TMDL programme flexibility to enhance stakeholder participation.
- Accomplish TMDL implementation through the state continuing planning process.
- Enhance opportunities for trading; i.e., PS:NPS trading ratio approaches 1:1.

If trading markets are designed for specific TMDLs, two fundamental forms of uncertainty will be behavioural and stochastic uncertainty. Specifically, if farmers are allowed to participate in a trading programme to meet TMDLs, will they participate? Will the allowable practices achieve the desired abatement? Moreover, how will the market and water quality respond to stochastic weather shocks? First, the question of farmer participation is fundamental to a point-nonpoint trading programme for TMDLs. Furthermore, trading programmes are relatively new for regulating water pollutants, and so a certain degree of reluctance on the part of regulators and sources is expected, because the economic and environmental outcomes are unclear. Clearly defined rules for participation, touching areas of uncertainty mentioned earlier, are essential to encourage farmer participation. For example, what is the unit of trade? Is it a pound of phosphorus prevented from reaching the edge-of-field, a drainage ditch, or the mouth of the stream? The questions of water quality and practice uncertainty are obviously interrelated and will affect the policy choice for meeting TMDLs. While biophysical models can generate good estimates of expected NPS abatement at the field's edge, the actual load of pollutant arriving at the water body is still uncertain. Trading programmes have dealt with this uncertainty by setting different trading ratios for different sources that reflect the uncertainty involved with the expected level of abatement (Malik et al. 1993). Such trading ratios are often required to provide "reasonable assurance" that the target level of reductions will be met. In practice, trading ratios serve as an explicit transaction cost, making it more expensive for PSs to purchase NPS source offsets.

What we might expect to see in practice is adaptive management, where the policy iterations based on observed ambient outcomes might result in different trading ratios among sources, different initial distributions of permits to sources, or imposition of locality constraints for trades to avoid hotspots. After implementing a trading programme for nutrients in a particular watershed, for example, the water quality manager might notice that not many trades are occurring and water quality is not improving. A water quality manager could then decrease the number of permits offered initially to sources and the trading ratio between point and nonpoint sources. Both of these actions would encourage more trades, and likely reductions in overall pollutant discharge. This example is reminiscent of the next case study we mention.

4.4.2 Tar-Pamlico

The Tar-Pamlico water quality plan is called a market, but it is not a market in the standard sense. Relatively few point sources contribute roughly 25 percent of the pollution and multiple nonpoint sources the remainder. It was thought that abatement costs would be relatively high for point sources and relatively lower for nonpoint. The programme establishes responsibility at the group level (a discharger Association for point sources, and the Agricultural Cost Share Programme for nonpoint sources) for reducing nutrient (nitrogen and phosphorus) loads to the major rivers in the basin. A discharge cap is set for the point sources as a whole. They can pay an offset fee for each mass unit of pollutant by which they, as a group, exceed the cap each year. The fee is set by the state (not by a market). The offset funds go to a voluntary agricultural cost share programme used to pay willing farmers 75 percent of the cost of installing nutrient-reducing best management practices (BMPs). A trading ratio of 2:1 is set for confined animal operations, and 3:1 for cropland BMPs.

Thus, the Tar-Pamlico programme establishes responsibility at the group as opposed to the individual, level and there are no individual polluter-level transactions. Initial Phase I requirement of 200,000 kg abatement was achieved by point sources. However, there have been no formalised trades. These reductions were accomplished through collective action on the part of the regulated point sources (Association) by installing more advanced pollution control technologies and making other changes in the treatment process. It is unclear if these reductions would not have occurred under a typical (non-tradable) quota reductions policy.

To the extent that the Association equalises marginal costs of abatement across point sources, the programme can be considered an informal version of point-point trading. Yet it is better described as tax on point sources that exceed their caps, the proceeds of which are applied to a more cost-effective method of achieving the reductions (agriculture). The use of average cost prices for point-nonpoint trades (set by the state), rather than marginal cost prices, is another deviation from a true trading programme.

One reason that no point-nonpoint trades have occurred may be that PSs have not yet exceeded their caps. When the programme started in 1989, the PS were required to fund an efficiency study on all of their facilities, and to implement the study's recommendations for optimising plant performance for nitrogen and phosphorus removal. This action, combined with installation of biological nutrient removal at a couple of the larger facilities as they underwent expansion, yielded sufficient reductions and allowed the Association to stay beneath its cap each year despite increases in flow. Consequently, the use of flexible technology requirements, as opposed to uniform requirements, has resulted in lower costs. While this has reduced PS discharges, the overall goal of the programme has not been achieved because of lack of progress on the nonpoint side, which remains strictly voluntary.

In short, the benefits of the Tar-Pamlico trading programme are difficult to assess. Many consider the programme a failure since there has yet to be point-nonpoint trades, which was the original intent. Nonpoint emissions have not been effectively incorporated into the trading programme due to high point/nonpoint trading ratios, average (not marginal) cost pricing, and non-binding emissions restrictions. However, the potential remains for the Tar-Pamlico trading programme to achieve the identified water-quality goals at least cost, though a number of changes will be required (Hoag and Hughes-Popp 1997). If the aggregate point-nonpoint loading restriction was set at the recommended 45 percent (North Carolina Department of Environment and Natural Resources 2001), there would be a much greater incentive for nonpoint-nonpoint, point-point, and point-nonpoint trades.

4.4.3 Pollution Trades in the Minnesota River Basin

After a yearlong study, a Minnesota Pollution Control Agency (MPCA) steering committee recommended the use of point-nonpoint pollution trading to help reduce pollution levels on the Minnesota River. In response to these recommendations, MPCA negotiated with Rahr Malting Company (Rahr) in 1997 and with Southern Minnesota Sugarbeet Cooperative (SMSC) in 1999 to start pollution trading in the Minnesota River Basin with NPS. This was done under the National Pollution Discharge Elimination System (NPDES) permits issued to the two PS. The pollutants being traded were essentially phosphorous and nitrogen although they were also translated into $CBOD_5$. The two PS with the help of MPCA had to bear the transaction costs of identifying NPS trading partners and ensuring that the pollution reduction practices were implemented. All trades had to meet four conditions: efficiency, equivalence, additionality, and accountability.

A mandatory trust fund was established by each of the two PS to assure the financial viability of the trading project. A trust fund board, composed of at least one local watershed manager, one government representative, and one local water resource organisation representative, had responsibility for managing the trust fund and approving trades. The minimum amount of the trust fund was specified in each permit ($250,000 for Rahr and $300,000 for SMSC).

The two trading projects employed trading ratios equal to or greater than 2:1, two units of NPS reduction for one unit of PS. The trading ratios were developed

1) to account for uncertainties in converting NPS loads into PS loads and 2) to provide additional environmental benefits on the River. The final trading ratios were the results of negotiations among the permit-holder, MPCA, and public participants. Every trade had to be verified by MPCA and annual reduction goals were outlined in each permit. MPCA had the right to revoke tradable credits based on inspection results. Annual reports were also required from the PS as specified in their permits.

In order to build its own wastewater treatment plant, Rahr needed to buy enough pollution abatement credits to offset the plant's potential wastewater discharge. It was able to do this through four trades with NPS. The NPS control practice included two trades converting farmland back to its original floodplain status with native grasses and trees to stabilise the soil and prevent flood scouring. The other two trades involved stream-bank stabilisation and livestock exclusion plus stream-bank erosion control. The trades were designed to reduce phosphorous and nitrogen losses into the River from the NPS.

An evaluation of the Rahr trades found them to be cost-effective when compared to alternative treatment costs (Fang and Easter 2003). Even when transaction costs raised the total outlay by 34.6 percent, the Rahr trades were still cost-effective. The cost-effectiveness increased significantly when the assumed life of the practices was increased from 10 to 20 years.

The second trading programme, with SMSC, was initiated as a means to offset the planned wastewater discharges from its proposed wastewater treatment plant. This was done by contracting with the cooperative's sugar beet growers to plant a spring cover crop as an erosion control BMP. The cover crop provides fields with some vegetative cover before the sugar beets emerge and when the potential for soil-eroding rainfall events is particularly high. Four or five weeks after the beets emerge, the cover crop is killed with post-emergence herbicides. For SMSC, the cover crop was the easiest way to obtain pollution discharge credits for the Minnesota River, because of the cooperation of its growers. During 2000 and 2001, 164 landowners planted spring cover crops on 35,839 acres of sugar beet land. Each grower received $2 per acre from SMSC to plant the cover crop at an average cost to the farmer of $6 per acre.

The SMSC programme did not turn out to be nearly as cost-effective as the Rahr programme (Fang and Easter 2003). It cost the SMSC growers $18-24 per pound of phosphorous reduced. This is higher than the cost of waste treatment plants ($4-18) and much higher than the cost of phosphorous reduction ($2-4 per pound) in the Rahr trades. In addition, the cover crop must be planted each year in contrast to the longer-lasting changes resulting from Rahr's trades. Both of these NPS estimates include transaction expenses, which raised their costs by 30-35 percent. On the positive side, a number of additional benefits included a reduction of soil erosion and an elimination of odours from the SMSC wastewater ponds which were no longer needed.

The two Minnesota River programmes illustrated several key problems with PS and NPS trades. First, is the uncertainty in the NPS trades (what should be the ratio between PS and NPS trades?). Second, the transaction costs (administrative) are likely to be high. About 65 percent of transaction costs occurred before the trades while 81 percent of total transaction costs were borne by MPCA. Finally, without any requirement to reduce discharge loads, NPS have limited incentive to participate in trades, which raises the transaction costs of obtaining willing trading partners. In reality, neither of these programmes can be considered a free-market exchange. The search for trading partners could have been expanded and lower-cost options found. In addition, MPCA closely controlled the process and bore a large share of the transaction costs. Yet it is possible to build on this experience and develop a more open trading system.

4.5 Conclusions

Conventional policies to manage water quality have met with varying degrees of success. It may be that innovative management policies that employ market mechanisms offer a solution to persistent water quality impairments. For example, in the Chesapeake Bay over the past 20 years, billions of dollars have been spent on mitigating the harmful effects of nitrogen and phosphorus discharge. Yet, according to data gathered between July 7 and 9, 2003, approximately 40 percent of the water in Chesapeake Bay had low dissolved oxygen levels – less than 5 mg/l – causing stressful conditions for many species of fish and shellfish (Chesapeake Bay Program 2004). Further, it is estimated that an additional $15 billion in Federal funds over the next six years will be needed to achieve water quality goals for the estuary (Chesapeake Bay Watershed Blue Ribbon Finance Panel 2005); while current Federal resources available for mitigation efforts stand at an estimated $230 million per year (Washington Post 2005). It is clear in this case, and in those of many other waterways, that current systems of regulating point sources through technology standards and relying on ad hoc or voluntary controls for nonpoint sources are not sufficient to meet water quality goals. Are markets the solution? Can the rest of the world gain insight from the United States experience so that they can essentially "leapfrog" from less efficient to more efficient water quality management?

Fees on discharge, though capable of achieving the same result as tradable permits, are politically unsavoury, sources incur abatement costs and are required to pay discharge fees. The information necessary to set fixed abatement standards for each source is expensive to obtain, functioning markets can provide this information free of charge. Moreover, tradable markets, as do effluent fees, encourage innovation across technologies and sources, so serving to reduce abatement costs over time. The market also allows those buying permits essentially to pay the sellers to reduce pollution. That said, a discharge trading system has a number of market frictions, or transaction costs, which must be addressed for a market to function and for water quality goals to be met.

It is not unrealistic to envision a permit-trading structure beginning as a bilateral mechanism, where perhaps only one of the sources is liable for reductions (e.g. Rahr Brewery). In such a case, a water quality management agency would be responsible for verifying that the proposed bilateral trades conform to established water quality standards. After adaptive management iterations, the advent of new sources, or the further degradation of the water resource, the bilateral trading structure could evolve as the number of potential permit buyers and sellers (and bilateral transactions) increase sufficiently to warrant the water quality manager setting up a clearinghouse. Under such a structure (perhaps based on a Nutrient-Net-Ebay type of setup), a water-quality manager could track and verify numerous permit trades based on a set of predetermined trading ratios (reflecting spatial and temporal damages or relative uncertainties). Lastly, after further adaptive management iterations and experience of the market participants, the clearinghouse could evolve into a pure exchange.

What conditions might facilitate the evolution of market mechanisms for water quality? Reiterating the King and Kuch (2003) suggestions, the number of potential buyers and sellers in a water-quality market needs to be sufficiently large. This can be accomplished by imposing tighter discharge limits on point sources (increasing demand for offsets) or broadening the geographic area for the water-quality market (increasing demand and supply). But in many cases nonpoint sources contribute the majority of the discharge, in which case it may be necessary to impose discharge limits on them (increasing demand and supply). In any case, an unbiased water quality agency is required to set-up the trading programme, establish the right to pollute, monitor permit trades and discharge levels, levy fines for non-compliance, and address physical and stochastic uncertainty through adaptive management. Monitoring and enforcement of permit holdings and discharge levels are required. They also may need to set trading ratios. Fees on excess discharge should reflect the opportunity cost of abatement, if due to stochastic influences such as weather. In order to discourage non-compliance, fees on excess discharge should exceed the opportunity cost of abatement. The costs of doing this can be passed along to taxpayers if the benefits of enhanced water quality outweigh the additional tax burden on society.

4.6 References

Arrow KJ, Hahn FH (1971) General Competitive Analysis. Holden-Day, San Francisco

Baumol WJ, Oates WE (1988) The Theory of Environmental Policy. Cambridge University Press, Melbourne, Australia

Chesapeake Bay Program (2004) Oxygen Levels in Chesapeake Bay. http://www.chesapeakebay.net/pubs/waterqualitycriteria/cost_backgrounder_final.PDF

Chesapeake Bay Watershed Blue Ribbon Finance Panel (2005) Saving a National Treasure: Financing the Cleanup of the Cheseapeake Bay. http://www.chesapeakebay.net/pubs/blueribbon/Blue_Ribbon_fullreport.PDF

Coase RH (1960) The Problem of Social Cost. Journal of Law and Economics 3: 1-44

Dales JH (1968) Pollution, Property, and Prices. University of Toronto Press, Toronto, Canada

Environomics (1999) A Summary of US Effluent Trading and Offset Projects, US Environmental Protection Agency Report. Office of Water, Washington, DC

Fang Feng, Easter KW (2003) Pollution Trading to Offset New Pollutant Loadings – A Case Study in the Minnesota River Basin. (American Agricultural Economics Association Annual Meeting, Montreal Canada)

Gleick P (1999) The Human Right to Water. Water Policy 1(5): 487-503

Gardiner R (2002) Freshwater: A Global Crisis of Water Security and Basic Water Provision. http://www.earthsummit2002.org/es/issues/Freshwater/freshwater.pdf

Hagem C, Westskog H (1998) The Design of a Dynamic Tradable Quota System under Market Imperfections. Journal of Environmental Economics and Management 36(1): 89-107

Hahn RW (1984) Market Power and Transferable Property Rights. QJE 99: 753-765

Hanley N, Shogren JF, White B (1997) Environmental Economics in Theory and Practice. Oxford University Press, New York

Hoag DL, Hughes-Popp JS (1997) Theory and Practice of Pollution Credit Trading in Water Quality Management. Review of Agricultural Economics 19(2): 252-262

Hoel M, Karp L (2001) Taxes and Quotas for a Stock Pollutant with Multiplicative Uncertainty. Journal of Public Economics 82(1): 91-114

Horan RD, Shortle JS, Abler DG (1998) Ambient Taxes When Polluters Have Multiple Choices. Journal of Environmental Economics and Management 36: 186-199

Johansson RC (2002) Watershed Nutrient Trading under Asymmetric Information. Agricultural and Resource Economics Review 31(2): 221-232

Johansson RC, Gowda PH, Mulla D, Dalzell B (2004) Metamodeling Phosphorus BMPs for Policy Use: A Frontier Approach. Agricultural Economics 30: 63-74

Kaplan JD, Howitt RE, Farzin YH (2003) An Information-Theoretical Analysis of Budget-Constrained Nonpoint Source Pollution Control. Journal of Environmental Economics and Management 46(1): 160-30

King DM, Kuch PJ (2003) Will Nutrient Credit Trading Ever Work? An Assessment of Supply and Demand Problems and Institutional Obstacles. Environmental Law Reporter 33: 10352-10368

Malik AS, Letson D, Crutchfield SR (1993) Point/Nonpoint Source Trading of Pollution Abatement: Choosing the Right Trading Ratio. Amer. J. Agri. Econ 75: 959-967

McCann (1997) Evaluating Transaction Costs as Alternative Policies to Reduce Agricultural Phosphorous Pollution in the Minnesota River. Ph.D. Thesis, Department of Applied Economics, University of Minnesota

McCann LJ, Easter KW (1999) Differences between Farmer and Agency Attitudes Regarding Policies to Reduce Phosphorus Pollution in the Minnesota River Basin. Review of Agricultural Economics 21(1): 189-207

MPCA (1997) Permit MN 0031917. Minnesota Pollution Control Agency Water Quality Division, Point Source Compliance Section, St. Paul, MN

Moledina AA, Coggins JS, Polasky S, Costello C (2003) Dynamic Environmental Policy with Strategic Firms: Prices Versus Quantities. Journal of Environmental Economics and Management 45(2S): 356-376

Montgomery WD (1972) Markets in Licenses and Efficient Pollution Control Programs. Journal of Economic Theory 5: 395-418

North Carolina Department of Environment and Natural Resources (2001) Frequently Asked Questions about the Tar-Pamlico Nutrient Trading Program. http://h2o.enr.state.nc.us/nps/FAQs9-01prn.pdf

O'Neil W, David M, Moore C, Joeres E (1983) Transferable Discharge Permits and Economic Efficiency: The Fox River. Journal of Environmental Economics and Management 10: 346-355

Segerson K (1988) Uncertainty and Incentives for Nonpoint Pollution Control. Journal of Environmental Economics and Management 15(1): 87-98

Smith RBW, Tomasi TD (1995) Transaction Costs and Agricultural Nonpoint-Source Water Pollution Control Policies. Journal of Agricultural and Resource Economics 20(2): 277-290

Stavins RN (1995) Transaction Costs and Tradable Permits. JEEM 29:133-148

Stavins RN (2003) Market-Based Environmental Policies: What Can We Learn form US Experience (and Related Research)? Resources for the Future Discussion Paper 03-43, Washington, DC (August)

Taff S, Senjem N (1996) Increasing Regulators' Confidence in Point-Nonpoint Trading Schemes. Water Resources Bulletin 32(6): 1187-1194

UNEP (1999) Global Environment Outlook 2000. United Nations Environment Programme, Earthscan Publications, London, UK

EPA (2002a) Total Maximum Daily Loads: National Overview. US Environmental Protection Agency. http://www.epa.gov/owow/tmdl/status.html

EPA (2002b) Status of TMDL/Watershed Rule. US Environmental Protection Agency. http://www.epa.gov/owow/tmdl/watershedrule/watershedrulefs.html

EPA (2002c) Trading: Case Studies. US Environmental Protection Agency. http://www.epa.gov/owow/watershed/hotlink.html

EPA (2004) Atlas of America's Polluted Waters. US Environmental Protection Agency. http://www.epa.gov/owow/tmdl/atlas/index.html

EPA (2005) Overview of Current Total Maximum Daily Load -TMDL- Program and Regulations. US Environmental Protection Agency. http://www.epa.gov/owow/tmdl/overviewfs.html

Washington Post (2005) States Seek U.S. Funds for Bay: Cleanup Far Behind Schedule. By David A. Fahrenthold (January 11, p B01)

Weitzman ML (1974) Prices vs. Quantities. The Review of Economic Studies 41(4): 477-491

Woodward RT (2003) Lessons about Effluent Trading from a Single Trade. Review of Agricultural Economics 25(1): 235-245

Woodward RT, Kaiser RA (2002) Market Structures for US Water Quality Trading. Review of Agricultural Economics 24(2): 366-383

World Bank (2003) World Development Indicators 2003. The World Bank Group, Washington, DC

World Resource Institute (WRI) (2004) NutrientNet. http://www.nutrientnet.org/prototype/html/index.html

Xepapadeas AP (1991) Environmental Policy under Imperfect Information: Incentives and Moral Hazard. Journal of Environmental Economics and Management 20:113-126

Integrated Water Quality Management in Brazil

Benedito Braga, Monica Porto and Luciano Meneses

5.1 Introduction

Integrated water resources management is a concept that has had an important appeal worldwide in the last decade. The concept of integrated water resource management (IWRM) has broadened substantially in recent years to emphasize the complexity of the sustainable use of water, given the multiplicity of its uses, its socio-economic importance, and the many forms of degradation that may affect this resource. This concept involves the consideration of interactions between soil and water, surface water and groundwater, water quantity and water quality, just to name a few. Integrated management of water quantity and quality is necessary in order to have a technical and institutional framework that will enable the restoration of aquatic ecosystems as well as the development of hydraulic infrastructure in a sustainable fashion. This is the first step towards more encompassing concept of IWRM involving social and political considerations.

This chapter describes the mechanisms that are needed to ensure the feasibility of management of water quality and quantity in a river basin. Emphasis is given to the application of economic instruments together with the traditional command and control mechanism as an alternative to enable a more efficient way of managing scarce resources. The latter has proven to be inefficient in less developed countries. Brazil, for instance, which has some of the most advanced environmental legislation in the world, has not been able, using only command and control, to reduce pollution in its urban rivers for the last 30 years. A combination of instruments is necessary to achieve sustainability and efficiency, including plans for river basins and the adoption of charging for the use of water as an economic instrument to ensure more efficient use. Control is necessary but it is not sufficient, unless implemented in conjunction with the aforementioned instruments.

Pollution control is definitely a major problem in developing countries. In Latin America, the average coverage for sewage collection is 77 percent. In Brazil, 76 percent of the population is served by a sewage collection system, but only 26 percent by waste treatment plants. The major barrier to redressing the balance is the magnitude of the necessary investments, which frequently exceed the financial capabilities of developing countries. In Brazil, for example, it has been estimated that investment in the range of $80 billion are necessary to solve the problem.

A different orientation in the way that pollution control is currently viewed may change this situation. In many developing countries, water scarcity is due to inefficient use and to pollution that results from uncontrolled effluent discharges. Pollution control in these cases goes far beyond environmental protection. It may represent a significant increase in water availability and, therefore, an important opportunity to provide for new users. Potential conflicts are reduced and an important opportunity for may arise for the renewal of regional development, generating more income and thus, more investment. Viewed in this manner, a new approach should be taken for water quality management.

The challenge then presented to decision-makers is how to use management instruments to solve both water quantity and quality issues in an integrated manner. The water resources management system and the environmental management system must work with compatible views on water uses, employing adequate and integrative instruments. This chapter presents an approach for the issue of water permits and for charging that integrates the water resources and environmental management systems, reducing both to a single parameter: the river flow. Water permits are awarded on the basis of the amount of flow allocated both for direct use or withdrawals and for pollution prevention. The user is charged for the river-flow utilisation in both cases.

5.2 Water Quality Management Framework

Many countries have chosen to have an environmental management system. In general these systems involve the participation of the organised civil society, local governments and users, and are designed to protect the environment against anthropocentrism. The system works in a reactive way, through the analysis of environmental impact assessment reports which show mitigation measures to compensate environmental damages. Management systems may be used to change this approach and to implement processes that will focus on prevention.

When considering water, its uses and related impacts, integrated management becomes essential. Brazil, for instance, has an environmental management system in place as well as a water resources management system. It is clear that the decisions, actions and all kind of procedures related to water uses must be taken in consideration of both systems. Such systems, for the most part, are linked by the quality aspects of the water, either because of flow alterations or impact on characteristics of water quality.

IWRM has an important role in bridging this gap that may be defined as the establishment of a water quality management effort. This effort must encompass legal aspects, institutional arrangements and financing mechanisms, to create a water quality management system. Water quality management systems are recognised as essential since it is impossible to implement any human activity, such as urban occupation or agricultural production, without causing stress to the environment. The management process aims to maintain the stress to a sustainable level. It searches for equilibrium between water and land uses, with an acceptable and realistic risk for environmental and changes.

Water quality management is no longer a subject that deals with technical matters only. It must include several other aspects, such as the economic impact on human activities, political negotiation and social analysis of the potential pitfalls that can result from environmental degradation or economic constraints. It is important to recognise that such management systems must allow several water uses while maintaining concurrent pollution generation within sustainable levels, as defined by environmental, economic, political and social variables. In addition, the management system must have clear and efficient decision mechanisms in order to be acceptable and operational.

A water quality management system should follow these guidelines (Larsen and Ipsen 1997):

- Use of preventive mechanisms. Remedial measures usually cost more and may even be less effective.

- Regulations and standards must be realistic to be achievable and enforceable. Unrealistic standards and non-enforceable regulations may do more harm than good, since they undermine trust in the rules of law. Laws and regulations should be formulated to match the level of economic and administrative capability of management institutions. To achieve improvements, they may be gradually tightened during the evolution of the processes, but lack of enforcement is one of the most common failures of pollution control systems in developing countries and its cause is usually regulation that is exceedingly restrictive, and thus unenforceable.

- Command and control mechanisms have to be balanced with economic instruments. The former are usually preferred because they offer a reasonable degree of predictability about pollution reduction. On the other hand, a major disadvantage of command and control system is its economic inefficiency and the administrative burden that it imposes on the controlling authority. A mixture of both approaches seems to be the most efficient way to control pollution.

- The polluter-pays-principle should be applied. This economic instrument implies that the costs of measures to prevent, control and reduce pollution must be taken by the polluter. Its aim of to induce a change in behaviour that will tend to decrease environmental damages.

- Mechanisms have to be established for cross-sectoral integration, ensuring coordination of pollution control efforts with water uses.

- Decisions must be implemented at the appropriate level, which can be defined as the level at which significant impacts are experienced. Decisions should be based on participation and decentralisation.

- Reliable information systems must be implemented and given open access to stimulate understanding and participation. The major challenge is how to produce information for different levels of stakeholders, from manager down to final user.

Water quality management is an ongoing process that will define:

- General policies for water uses, to combine conservation and/or restoration of the water bodies, economic development of the region and improvement of the living conditions in the basin. Priorities among these three aspects must be discussed and decided.

- Water quality objectives, or designated uses, that will require a certain condition of water quality to be sustained.
- A strategy to reach the defined objectives that balances the wishes of the community with financial capability and an acceptable timeframe.

There are several reasons why the integrated approach is essential to systems of water quality management. Such systems require a multidisciplinary view. Any acceptable solution depends on a global view of the basin as well as a focal view of the polluter. Different management instruments will be used for the environmental management system and the water resources management system. Last, but not least, general regulations tend to be defined by the central government, whereas decisions and implementation are done locally.

5.3 Institutional Arrangements

Articulation of the water resources management system with the environmental sector is fundamental. Water cannot be regarded as "just another" natural resource. Its importance to human and animal life, its role as a development engine (hydropower, navigation, aquaculture, irrigation), its importance in public health, and sometimes its role as a national security issue in countries such as Egypt, Turkey, Syria, Palestine and Israel, justifies the existence of a specific system for its management. In most countries, water is a public resource and thus requires adequate regulation to ensure access to all users. In Brazil, the National Water Resources Management System was established in 1997. It comprises the following institutions:

1. National Water Resources Council,
2. National Water Agency (ANA),
3. State water resources councils,
4. River basin committees,
5. Agencies at the federal, state and municipal levels, related to water management, and
6. Basin agencies, the executive office of the river basin committees.

The main tasks of this set of institutions are to implement the water resources policy, to coordinate integrated management and to make the water charging system effective. This policy is executed by the National Water Agency (ANA).

Categories of members forming the National Water Resources Council (NWRC) are (i) representatives of the federal government, from ministries with some level of jurisdiction over water and related issues; (ii) a representative from the National Water Agency; (iii) representatives designated by the state water resources councils; (iv) representatives of the stakeholders; and (v) representatives of non-governmental organisations involved with water resources management or water use. Among the main responsibilities of the NWRC are: (a) to arbitrate, in the final instance, conflicts between states; and (b) to approve guidelines regarding the permit system for withdrawals and water use, and also for the implementation of bulk water charges. The executive secretariat of the Council is under the responsibility of the Office of Water Resources, Ministry of Environment.

In 2000, the National Water Agency (ANA) was established. ANA is the executive branch of the system, responsible for the implementation of the water resources policy, and all related technical tasks. The tasks include the National Water Plan, the permit and charging system for federal rivers, and establishment of an information system. ANA is the regulator. Since ANA is responsible for issuing water-use permits, this is one of the regulatory mechanisms that ensure adequate use of water in terms of an overall basin perspective. Water pricing is another regulatory mechanism authorised by the water law. ANA can also establish the basis for implementation of rational use and conservation programmes, especially areas under water stress.

A river basin committee may be established in states that have the necessary legal provisions and in federal rivers. The underlying principle for such committees is that government, stakeholders and the civil society decide how to: (i) allocate water; (ii) implement new development projects; (iii) arbitrate conflicts among stakeholders, and (iv) impose pollution control restrictions; among other relevant issues related to water resources use and/or management. The basin agencies were designed to perform all the executive work related to water management in each basin. The funds for their operation should be provided through the collection of bulk water fees. The water agency is responsible for all the technical work required to manage water resources locally. The supply of expertise i.e., for data base management, hydrologic studies to assess water availability, for ensuring adequate water withdrawal decisions, for assessing and evaluating new water resources projects, as well as providing technical support to the committee on any other technical issue, are all responsibility of the Agency.

This system must work together with the National Environmental System, which was established in 1981. The latter system's main instrument of operation is the environmental permit, which aims to control environmental impacts that may be caused by potentially damaging activities. This is typically a command and control mechanism and, as noted earlier, of limited efficacy in developing countries such as Brazil. Currently, the Ministry of Environment operates the system but the National Environmental Council and several agencies at state level are in charge of issuing the permits.

The integration of both systems may create synergy because it will allow the integration of command and control mechanisms (water use permits and environmental permits) together with economic instruments (water pricing) under the framework of planning and management (basin plans and information systems). It is clear that, where water is concerned, the management framework must encompass both systems, bridging the usual gap between decisions on water quantity and water quality. The result is expected to be a major contribution to pollution reduction and control.

5.4 Integrated Water Management Instruments

The National Water Resources Policy offers several instruments that are needed for the implementation of IWRM. Nevertheless, correct procedures have to be established in order to apply them in an integrated manner. The effectiveness of these instruments will depend on the capacity to implement them in accordance with their objectives. The instruments are: water resources plans; definition of water quality objectives through classification of water bodies in accordance with designated uses; the permit system for both withdrawal and discharges; charging for water use, and the information system. All are means of integration. Integration of quantitative and qualitative aspects, according to each instrument, will be discussed next. The permit system and the charging processes will be explained in detail, mostly because they represent the major challenges to the system and due to their potential for achieving IWRM targets.

5.4.1 Water Resources Plans

River basin plans are the first instruments to be developed for the implementation of basin management systems. They define priority uses, and thus water availability, both in quantitative and qualitative terms, matching it to expected demand. Other factors of integration, such as land use and occupancy, economic growth, social development must also be taken into account. The plans define the investment needed to fulfil the expected demand, i.e. the goals to be achieved in terms of both quantity and quality. As previously noted, the policies for water uses and the respective expected quality must be defined as a part of these instruments. The plans are the instrument through policies, water quality objectives and restoration or conservation targets are established.

5.4.2 Water Quality Objectives and Designated Applications

The definition of water quality objectives to protect designated applications is the tool that integrates quantitative and qualitative aspects of water management. They are the water quality goals that should be achieved to enable expected applications. The definition of water quality objectives to support designated applications in the basin has a direct impact on the investments and economic growth of the region. If the objectives are too restrictive, major investments will be needed to control pollution. If they are less restrictive, or even loose, the need for investment is lessened but the pollution levels will probably reduce water availability and slow down, or even prevent, economic activities. Care must, therefore, be taken to achieve equilibrium between water quality objectives and the required investment in order to maintain adequate economic development with achievable targets for restoration of water quality.

In Brazil, water quality objectives are classified under a system that seeks to ensure that quality is comparable with the most restrictive uses. In order to manage a basin in accordance with the established water quality objectives, it is necessary to take not only corrective measures but also preventive procedures which, in the long run, will help to reduce the costs of managing pollution. Classification is, thus, the way to plan water quality management in basins. This is defined under the National Environmental System, with emphasis on the integration upon which water management is based.

For each river reach, the basin committee decides on a set of water quality objectives and select the corresponding classifications. For freshwater, Brazilian law (Resolução CONAMA No. 20) defines five classes of designated applications: (i) Special class, to be used for water supply and environmental protection, (ii) Class 1, for water supply, recreation, aquaculture, irrigation of more restrictive crops and environmental protection, (iii) Class 2, for water supply, recreation, irrigation and aquaculture, (iv) Class 3, for water supply with advanced treatment systems, irrigation of trees or fodder crops, (v) Class 4 for less restrictive uses. The basin plan must decide on the water classification of the bodies within its limits, and it forms the basis for the other regulatory instruments such as permits and charging.

5.4.3 Permit System

In most countries, water is considered to be a public property. Hence, its use has to be authorised by the government institution in charge of water management, usually, the ministries of water resources and environment. The various applications (urban, irrigation, hydropower generation, navigation, environmental protection, recreation, etc.) could be in conflict with environmental protection. Therefore, the management of water resources is mandatory, particularly if the economic, social, and environmental water demands are to be sustainable.

Water permits are necessary to ensure a legal mechanism for water allocation that meets the water quality and quantity needs of the various users, and allows for protection of supplies. Permit is an essential instrument for water management, because of its legal, economic, and technical implications. If well organised, these can contribute to the rationalisation of water use. However, to make it feasible, the individual and cumulative impacts of all users must be known in relation to both quantity and quality. This allows for multiple applications of the water resources.

In Brazil, the Water Law 9433-97 has established water permits as one of the instruments for water resources management. It constitutes a central element in control, with the aim of guaranteeing public access to water. In addition, Law 9984-00 established the National Water Agency (ANA), whose responsibilities include implementation of the National Water Resources Policy, and authorising the use of the waters of the federal rivers. It is important to note that water permits are not transferable. Water uses which currently require licensing are: i) abstractions and outflows which alter the regime, quality or quantity of water in the water body; ii) hydroelectric works; and iii) groundwater from wells for consumption or production.

The need to obtain a permit to discharge liquid wastes into surface water bodies is an acknowledgement of the fact that it is impossible to eliminate all pollutant discharges into the environment. Discharge will always be necessary, and therefore must be controlled. The quantity of pollutants, whose discharge may be authorised, must be such that the receiving water body retains a water quality consistent with its class.

Brazilian water law allows the definition of insignificant use, i.e., that which does not significantly affect the regime, quantity or quality of water. This definition varies between watersheds, mainly due to the existing water supply and demand. For example, the Paraiba River Basin Committee established a limit of one litre per second as insignificant use (below this authorisation is not required). All authorisations are required to meet the river classification criteria of quantity and quality, as established in the river basin plans. In 2001, the National Water Resources Council passed a resolution that aimed to reduce red tape and encourage transparency in the permitting process. The permitting authority, for example, has assured public access to the criteria and guidelines.

The water agencies, as established under Law 9433, do not have permitting authority, but may receive requests and analyse them before forwarding them to the one that does. This is expected to enable the local water agencies to build up knowledge base which should allow it to conduct proper technical analyses in the future. This capacity building should enable the agencies to make appropriate decisions in the future.

The water permitting process consists of at least three steps:

- Technical evaluation;
- Evaluation of the enterprise; and
- Legal analysis.

The technical evaluation consists of verification of the availability of water from a water body, i.e. whether the flow rate that is being requested can be supplied, both in terms of quantity and quality. The impact of the request on the watershed is also analysed. The minimum information required for technical analysis of the water permit includes the identification and characterisation of the intended use; required quantity; its seasonal variation; effluent characteristics (quantity and quality), and local water availability, as well as previous water demands on the basin. The technical analyses are to take into account the guidelines for the river basin plan, which provides the relevant essential information.

With reference to the analysis of the permit requests, flow abstraction and effluent discharges are studied, including consideration of the physical-chemical properties of the latter. Analysis also considers the type and magnitude of the request, avoiding overestimates by users who intend to speculate with the surplus, or underestimates where users intend to pay less in tariffs. Legal analysis examines the documentation and legal status of the permit request.

5.5 Water Permits for Quantity and Quality

In order to integrate within an authorisation system both water abstraction and effluent discharges, a common unit for measurement is necessary. This unit should consider both processes in terms of volumes and flows. For the abstractions, this is obvious since the user asks for a permit to withdraw a monthly or yearly flow volume. In the case of effluent discharges, the volume refers to the flow that is needed to dilute, assimilate or transport the effluent in order to maintain water quality objectives. There are several advantages in using this methodology. One is the estimation of water availability, or the balance between supply and demand, using a common unit that helps ordinary water users to understand the processes. Conflicts are easier to resolve when simple concepts are used.

5.5.1 Estimating Dilution River Flow

A given user may affect the quality or quantity, or both, of the water body. Depending upon the quantity of effluent as well as its composition and concentration, the discharge may or may not be considered harmful to the environment and to other users. Quality and quantity impacts to the water body need to be known through time and river reaches, and for each quality parameter. Once individual impacts have been established, the cumulative effect on water bodies is estimated.

The quality impact on the water body is transformed to its quantity equivalent. This procedure significantly facilitates the overall analysis. For effluent discharged in a water body, the final concentration is given by the mass balance equation:

$$C_{mixture} = \frac{C_r.Q_r + C_{ef}.Q_{ef}}{Q_r + Q_{ef}} \qquad (1)$$

C_r = where, concentration of a given quality parameter in the water body,
Q_r = water body flow rate,
C_{ef}^{*} = concentration of a given quality parameter in the effluent discharge,
Q_{ef} = effluent discharge, and
$C_{mixture}$ = concentration of a given parameter in the mixture.

(*) If C_{efl} is not available, it can be estimated by
$C_{efl} = C_{tip} \times (1 - \alpha\beta)$, onde
C_{tip} = Standard concentration for the pollutant based on type and pre-treatment concentration
α = ratio of treated effluent volume over total effluent volume
β = treatment efficiency (0.90 for secondary treatment)

The necessary dilution or assimilation flow is based on the following equation (Kelman 1997):

$$Qdil = Qef . \frac{(Cef - Cperm)}{(Cperm - Cnat)}$$

(2)

where,
$Qdil$ = dilution or assimilation flow rate for a given quality parameter,
Qef = flow rate of the effluent containing the given quality parameter,
Cef = concentration of the water quality parameter of the effluent, and
$Cperm$ = maximum permissible concentration of quality parameter in the receiving water body.

The dilution flow rate (Qdil) is the flow necessary to dilute a certain concentration (Cef) of a given parameter, such that the resulting concentration (Cmixture) becomes equal to the permissible concentration (Cperm). The permissible concentration in the receiving water body is that indicated by the river category (designated water use).

It is assumed that the receiving water body is in a condition of natural concentration (Cnat). For example, a relatively clean river has a BOD of about 1 mg/l (from decaying organic matter of leaves, fish etc) (von Sperling 1998). For parameters such as phenol, mercury and arsenic, natural concentration in the water body is assumed to be zero.

The natural condition of the water body is used instead of the actual concentration for the following reasons:

1. Individual impact of each user can be calculated in absolute terms, without the interference of others.
2. In the case of water bodies that are being reclaimed, the equation would yield a negative result, indicating lack of dilution water, i.e. a condition which is influenced by existing use that would hamper the licensing of new applications until reclamation is complete.
3. Two water users releasing effluents with the same characteristics, both in terms of quantity and quality, will be treated differently if their effluents are to be released over different periods, or at different points in the basin. For example, if one of the users releases effluents five years after a similar user, the dilution flow of the former would be higher, due to pre-existing conditions.

The result of equation (2) is a water body flow, named dilution flow (Qdil), which is "virtually" apportioned by the user, in order to dilute the effluent parameter. This flow propagates downstream, and can have its magnitude increased, decreased or maintained, depending:

1. Whether the quality parameter being diluted is conservative or non-conservative.
2. Whether the permissible concentration (Cperm) of the parameter in the downstream reaches will or will not experience changes.

Effluent flow is added to the dilution flow calculated by equation (2) resulting in the mixture flow rate. The concentration of the latter must not exceed the permissible concentration of the parameter, *Cperm*.

In the case of non-conservative effluent flows, such as BOD, the resulting concentration in the mixture flow (Cmix) will decrease with time and the river reach, as organic matter decomposes.

In the mixture flow of a given parameter no other dilution of the same parameter may occur. This flow is, therefore, referred to as unavailable flow (Qunav).

The unavailable flow (*Qunav*) at the discharging point is given by:

$$Qunav_1 = Qdil + Qef \tag{3}$$

However, this flow could be used to dilute other parameters, or be abstracted.

In order to maintain water quality, the water balance is calculated by quantifying, in all reaches, the total unavailable flow for each quality parameter, decaying or not, from the different user discharges. That balance has to consider all abstractions (i.e., quality demands) so that the remaining flow (*Qrem*) can be estimated. If the unavailable flow in any given month or reach is greater than the remaining flow (*Qunav* > *Qrem*), it means that the flow is insufficient to dilute the effluents and maintain the water body in the desired state or category.

The idea is to use the volume balance concept, both to represent the relative water quantity and quality demands, warranting a "space" within the water body sufficient to maintain quality, even after the effluent discharges. This is particularly valid for organic matter and nutrient-rich effluents, since the maintenance of low concentrations of these parameters in the water, despite large volumes, is frequently desired. However, for those pollutants produced in small quantities, or even those that are toxic or cumulative, this process may not be the most efficient. Different licensing approaches, such as environmental licensing, would be required to ensure the avoidance of environmental damage.

5.5.2 Pollution Abatement Pact

Water quality is affected in many ways by pollution, mainly in rivers that cross large urban areas, or in intensively used basins.

The main reason for such impacts is that there is not enough water for all applications if the river's support capacity is to be preserved. In order to regularise existing applications and control pollution, an agenda has to be established to maintain water quality objectives that will support designated uses. This would require each user, within a given period, to adopt progressive control measures that allow for the reduction of environmental impact and meet the respective water standards.

Figure 5.1 schematically represents a user pact for BOD pollution control where levels of tolerable pollution values during different periods are given. These values are gradually reduced so that the water quality goals are met. This gradual reduction in the tolerable limits has a direct relationship with the dilution flow calculations used for issuing permits. When applied to a situation like Figure 5.1, equation 2 indicates that if the user fails to reduce effluent discharge or concentration, the dilution flow, as estimated at the beginning, will increase whenever the permissible values (*Cperm*) are reduced.

Once the pollution reduction pact has been established and, at the same time, a water tariff on the unavailable flows (*Qunav*) is defined, the user who fails to comply will be paying more for discharging the same amount of pollutants. Depending upon the tariff levels, the user will notice that it is cheaper to treat effluents than to pay for the discharges. Caution must be taken in applying this type of approach in order to avoid excessive economic impacts to specific water users or group of users.

This water permit system is currently in use at ANA. Efforts are being made to involve the Environmental Agency of the Ministry of the Environment (IBAMA) into the process. It is important that the water user need go to only one institutions in order to request a permit from the federal government. The idea is that all issues related to environmental licensing should be handled by the water resources management agency, either at national or at state level.

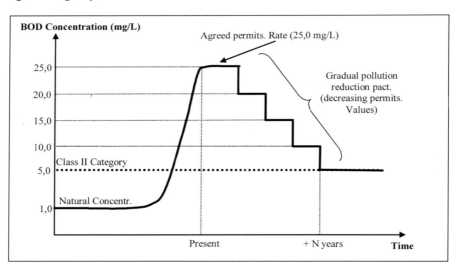

Fig. 5.1. Basin committee pollution pact

5.5.3 Charging for Water Use

Water has an economic value. The expression refers only to the use of the natural resource, and not the related services. In France, charges for water use have been in place for more than three decades. The system appears to be effective in controlling water shortages, and promoting sustainability and efficient operation of the infrastructure (Barth et al. 1987). More recently, Germany's Water Law introduced payment of the common control costs and Mexican Water Law charges for exploitation and use of surface and groundwater (Solanes and Gonzalez-Villarreal 1999). In Brazil, several states are also considering the possibility of introducing laws and regulations that would authorise charging. The main objectives of charging are to:

- Manage increasing demands and to promote efficient water use;
- Redistribute costs in a more equitable way;
- Emphasise the importance of water as an input to production and, as such, its economic importance; and
- Provide necessary funds for adequate operation and maintenance of existing systems and to implement new projects. It may also contribute to environmental conservation and restoration.

Experience shows that, in order to be effective, the revenues collected from bulk water fees must be invested in the basin where they are generated. This is the single most efficient way to increase willingness to pay. Otherwise, stakeholders will perceive the payments as taxes, a perception that will be very difficult to overcome. Agricultural sector will be most affected by the implementation of a pricing system, in spite of being the largest water user in the country. Global markets, fierce competition, and fluctuations in the price of agricultural products represent additional obstacles.

Together with a permit system, charging for water use could contribute to optimum allocation. The volumes granted to dilute and/or transport effluents are subjected to billing, once again emphasising the possibility of integrating management. Charging thus ensures a criterion of equity. In any discharge, a given volume of water will always be allocated for dilution, unless the effluent is already at levels of concentration below the limit for its class. For conservative pollutants, this is the equivalent of consumption, since this volume of dilution will always be necessary. For non-conservative pollutants, this volume will be gradually released as the pollution is degraded. What is being charged for, both in consumptive uses and in dilution, is the use of the same asset, either in the form of withdrawals from streamflow, or in the form of its use in the channel. Therefore, the user is left to decide whether to treat effluents to a standard above that required by the license, or pay for volumes necessary for dilution that which will maintain the classes of use of that body of water.

Another interesting concept is that the maximum concentration of each pollutant can be established by the basin committee. In accordance with the criterion that the necessary, volume of dilution and/or transport is that which allows the class of the receiving body to remain the same downstream from the discharge point, it is assumed that the class in the reach considered has already been legally established. If water quality objectives can only be supported by executing the investment programmes in the basins, and these are supported by the charging itself, the solution is in the institutional area: the basin committees must establish goals and quality objectives in accordance with the funds available for investment. Thus, at a given moment, charging would be based on the dilution volume needed to enforce the current water quality objectives. In the future, the volumes would be determined as a function of more restrictive objectives, established at each of the stages agreed to in the basin plan. As objectives become more restrictive, it would be necessary to permit increasing volumes of dilution, for the same volume and concentration of effluents. In this case, either the permitted "polluter" would have to increase the efficiency of treatment systems, which would be desirable from the environmental standpoint, or implement appropriate reuse practices to reduce volumes. This position is a positive outcome, since it will encourage the committees to establish the most restrictive goals possible, consistent with the main uses of the land and socio-economic and cultural characteristics of the basin.

The formula for estimating the charges for water abstraction, consumption and effluent discharge can be written as:

$$A = Q_{abs} \times P_{abs} + Q_{cons} \times P_{cons} + Q_{unv} \times P_{ef} \tag{4}$$

where,
A = monthly payment (Brazilian reais/month),
P_{abs}, P_{cons}, P_{ef} = public prices defined by the basin committee for abstraction, consumption and effluent dilution (R\$/m³),
Q_{abs} = monthly flow abstraction (m³/month),
Q_{cons} = monthly consumption of flow (m³/month), and
Q_{unv} = monthly flow required for effluent dilution (Eq. 3) (m³/month)

This is an evolution of the formula in use in the Paraiba River Basin (Braga et al. 2003), which operates with effluent discharges instead of pollutant loads. However, it does not allow for a differentiation in the unit prices for effluent discharges as presented in Equation 4. Efforts are now being made to provide the basin committee with this new approach for possible future use.

5.5.4 Water Resources Information Systems

It is not possible to use this system without appropriate information to support the decision-making process. Here again, integration is both an opportunity and a need. The preparation of the basin plans, decisions regarding classification, and sustainability of the permit and charging processes require information for proper operation and management. Water bodies needs to be extensively and regularly monitored, both terms of in space and time, in order to provide necessary information to the system. Accordingly, special attention must be given to water quality monitoring networks that are very scarce in the country. Networks are needed to identify the water quality trends, check on the effectiveness of the classification, surveillance requirements and detection of accidents. Data required for efficient water quality management includes list of polluters, biological monitoring, and periodical evaluation of the ecosystems.

5.6 Conclusions

The integration of management systems for water resources and the environment can be achieved through an adequate permit and charging process. Information management systems also play an important role, together with river basin plans. The latter provide the necessary information for supply and demand, both quantitative and qualitative. This allows the classification of watercourses according to their predominant use, and subsequently the limits for issuing permits for effluent discharge.

The water permit methodology described in this chapter is being applied to federal rivers in Brazil. The system has been automated. At present, the average time required for issuing a permit for federal rivers is of the order of three to four weeks. In the past, issuing a permit often took several months. The methodology proposed allows for the joint handling of water quantity and quality permits.

Charging for water is now a reality in the Paraiba River Basin. A new methodology has been proposed that will allow the consideration of different unit prices for quantity and quality issues. The river basin committee is expected take this proposition into consideration in the near future.

5.7 References

Barth FT, Pompeu CT, Fill HD, Tucci CEM, Kelman J, Braga BPF Jr (1987) Modelos para Gerenciamento de Recursos Hídricos. Editora Nobel/ABRH, São Paulo, Brazil

Braga BPF, Strauss C, Paiva F (2005) Water Charges: Paying for the Commons in Brazil. International Journal of Water Resources Development 21(1): 119-132

CONAMA. Resolução Conama No. 20 (1986). Conselho Nacional de Meio Ambiente, Brazil

Kelman J (1997) Gerenciamento de Recursos Hídricos: Outorga e Cobrança. In: Proc. XII Simpósio Brasileiro de Recursos Hídrico, Associação Brasileira de Recursos Hídricos, Vitória

Larsen H, Ipsen NH (1997) Framework for Water Pollution Control. In: Helmer R, Hespanhol I (eds) Water Pollution Control. E&FN Spon, London, pp 20-45

Solanes M, Gonzalez-Villareal F (1999) The Dublin principles for water as reflected in a comparative assessment of institutional and legal arrangements for integrated water resources management. GWP/TAC Background Papers, No. 3, GWP/SIDA, Stockholm, Sweden

Sperling von M (1998) Introdução à Qualidade das Águas e ao Tratamento de Esgotos. 2 edn, Belo Horizonte: Departamento de Engenharia Sanitária e Ambiental, Universidade Federal de Minas Gerais, Belo Horizonte, Brasil

Institutional Aspects of Water Quality Management in Brazil

Raymundo Garrido

6.1. Introduction

Discussion of the institutional aspects of water resources management is one of the most controversial issues in Brazil today. This is a natural progression in terms of advances in the management of water use in a society that for many years has been improving its environmental management practices.

It is likely that there will be a few difficulties of an institutional nature because, before the National Water Resource Policy was issued, the debate on environmental preservation included the consideration of ecosystem quality, within which the water quality of rivers, lakes and aquifers played an important role.

When the debate on water resource management began to gather momentum, it was realised that completing the task would be impossible, if attention was given only to the management of water quantity. Between 1978[1] and the beginning of the 1990s, the discussion about Brazilian watershed management focused on the study of outflows and their allocation to competing users. The quality issue was considered to be of secondary importance, and little attention was given to groundwater.

The creation in 1995 of the Water Resources Secretariat within the Ministry of the Environment, and the promulgation of the Federal Law number 9433 on January 8, 1997, followed by the establishment of the National Water Agency (ANA) three and a half years later, changed the panorama of water resources management in Brazil considerably. All these developments contributed to significant advances in the process, including a serious debate on how to solve the problem of water supply through conjunctive use of surface and groundwater sources.

The most important discussions were on the water quality issues and the roles of each of the institutions involved in water resource management. Accordingly, water quality issues, including decisions on what should be done institutionally, by which institutions and what management tools need to be used, became critical questions. Institutional and organisational aspects of water management became important issues for discussion.

The current thinking is that getting private sector financing for the sanitation sector of Brazil is probably the best and wisest way of solving the problem of poor

---

[1] The actual water resource management model in Brazil started in 1978, with the creation of the Executive and Integrated Watershed Committees, CEEIBH, when the first basin committees were established.

water quality, which has continued to be a national problem over the past few decades. Unless continuing water quality degradation is seriously tackled, and the recent trends are reversed, the nation will face serious health and environmental problems in the coming years.

Institutions, in the context of the present chapter, mean the entities that are associated with water management. Besides legal texts, existing markets, etc., one can find the following players: federal, state, municipal governments and non-government institutions, private entities, legislators and managers and a series of other elements, such as policy tools that play a direct or indirect role in the overall management of water resources, or contribute to it in some way.

Although the 1934 Brazilian Water Code referred to the quality issue, and explicitly stated that anyone who degraded a river or lake would be required to return it to its original state, it is fair to note that limited attention was paid to water quality only after the urbanisation process accelerated during the second half of the 1960s, as a result of the rural exodus to the urban centres.

During the first half of the 20th century, most of the population in Brazil (80 percent) was rural. Continuing rural exodus resulted in the formation of an urban majority. At present, 80 percent of the population live in cities and only 20 percent in the countryside.

Urban sprawls grew and this growth exploded with the creation of a housing bank, BNH, which conducted an intensive house-building policy in the country in an effort to offset a housing deficit sparked by a demographic explosion[2].

In spite of investing in sanitation and financing water supply and sewerage networks, efforts by BNH were simply inadequate to solve the domestic sewerage problem, and in particular the wastewater treatment problem. This includes the phase of the National Base Sanitation Plan (PLANASA) that was inaugurated at the beginning of the 1970s, and which had as its nerve centres state water supply and sanitation companies[3].

The industrial development of the country contributed to further water pollution. At present, industrial concerns have become more sensitive to the need for environmental conservation. Even though, as a general rule, industrial pollution is less than before, industry continues to be an important source of pollution, especially in the most developed and urbanised parts of the country in the south and southwest.

It is a fact that water pollution in Brazil is primarily due to lack of urban effluent treatment. Brazil belongs to the First World in terms of availability of clean drinking water. It is a Second World country in terms of urban sewage collection services. However, it is a Third World country when provision of urban effluent treatment services is considered. Unfortunately, there is not even one river, lake or aquifer near a large- or medium-size city in Brazil that is not polluted because of inadequate treatment of wastewater.

---

[2] The population of Brazil in 1970 was 70 million. Today it is about 184 million.

[3] The role of the autonomous municipal water and sewage services should not be ignored. They still receive technical, financial and operational support from the National Health Foundation, FUNASA.

The focus of this chapter, therefore, is to analyse the structural reasons for the deterioration in water quality in Brazilian watersheds, and to encourage discussions as to how institutions can be further strengthened so that quality of water in rivers, lakes and aquifers can be improved cost-effectively within a reasonable timeframe. An attempt will be made to draw some lessons which may contribute to the formulation of more appropriate public policies that could contribute to the prevention of the deterioration of the water quality of the country.

6.2 Institutional Arrangements for Sanitation Sector

It has been noted earlier that lack of urban effluent treatment is the main cause of water contamination in Brazil. If actions are to be taken to correct this unacceptable situation, the whole issue of urban wastewater treatment and disposal must be considered in its totality. Within the context of water quality management, it means that the sanitation sector is the problem and, at the same time, is the solution to the problem. The problem, historically, has been institutional, and has been primarily due to shortage of financial resources and lack of political will.

Following some comments on the recent development in the areas of sanitation, a brief plan will be presented which could contribute to finding a solution to the water quality problem in terms of institutional arrangements.

At the beginning of the 20th century, the administration of sanitation services in Brazil was handled by private concessionaires. The Federal Constitution of 1934 stipulated that municipalities were to provide public services. During the 1940s, the federal government decided to strengthen the sector by creating the National Works and Sanitation Department (DNOS) and the Special Public Health Service (SESP)[4].

During the 1960s and the 1970s, mixed economy associations were created in the states and were administered autonomously. For example, in the northeast, SUDENE[5] supported these companies with large investments, and also established two regional enterprises, Water and Sewage Company in the northeast (CAENE)[6], and Sounding and Drilling Company (CONESP), whose task was to pump water from groundwater sources.

---

[4] In spite of the important services rendered by DNOS to the country, it was terminated by the Collor Government in 1990, and SESP, which was later called SESP Foundation, is now called the National Health Foundation, FUNASA.

[5] The Northeast Development Superintendency, established in 1959, had the important role of reducing the disparities within the Northeast Region. The institution was terminated during the second term of President Fernando Henrique Cardoso. Its role was taken over by the Northeast Development Agency, ADENE, whose actions have not had much impact until now.

[6] The role of CAENE was diluted because the state companies were strong, as they are until today, and they received their political authority direct from the governor of each state. CAENE terminated its activities in 1972.

In 1961, the Inter-American Development Bank (IDB) signed agreement with the regional banks of the Brazilian states and with the sanitation companies of these states. The aim was to support investment in the cities of Belém, Natal, Recife, Rio de Janeiro, Salvador and São Luís. This support helped the sanitation sector to gather momentum, since they required the state sanitation companies to take on this vital role and to become more efficient in order to ensure that an appropriate return on the capital investment can be made.

In addition to the efforts of SUDENE and IDB, a system to finance the sanitation sector was developed through the National Housing Bank (BNH). Simultaneously, the administrative apparatus of the State was reorganised and consolidation took place under the decree number 200 of 1967. As a result, municipalities were removed from the process.

The BNH then absorbed a fund, established in 1967, to finance the sanitation sector. The fund was created through a cooperation agreement, signed in 1965 by Brazil and the United States, and was managed by DNOS.

The PLANASA, with the support of BNH, started a revolution in the Brazilian sanitation sector, significantly increasing service coverage, as can be seen from Table 6.1. The water supply and sanitation service coverage index is available until the year 2000.

Table 6.1. Evolution of the water and sewage coverage services in Brazil (Percent)

Indexes	1970	1980	1990	2000
Water				
Urban houses, distribution networks	60.5	79.2	86.3	89.8
Rural houses, distribution networks	2.6	5.0	9.3	18.1
Sanitary Sewage				
Urban houses, collection networks	22.2	37.0	47.9	56.0
Urban houses, cesspools	25.3	22.9	20.9	16.0
Rural houses, collection networks	0.45	1.4	3.7	3.3

Source: Urban Development Secretariat of the Presidency of the Republic, SEDU/PR, Brasilia, 2002

The PLANASA was responsible for sanitation between 1971 and 1989, when the National Urbanisation Programme (PRONURB) started. It ended in 1994. After PRONURB, sanitation in Brazil went through another period of relative neglect until it reached its present condition, which is one of total lack of definition in terms of institution. Consequently, the sector has lacked the necessary investments, and water quality improvements have been slow.

This period was called the pro-sanitation period. Its short life lasted from 1995 to 1998, when the programmed investments at national level were stopped. Later on, apart from the efficient Catchment Restoration Programme (PRODES), conceived and implemented by the National Water Agency (ANA), only a few isolated state programmes were completed.

A summary of the main features of these phases and their main difficulties are shown in Table 6.2.

Table 6.2. Phases of sanitation actions in Brazil

Main Programme	Main features and constraints	Water Investment ($bl)	Period	Sanitation Investment ($bl)	Total Investment ($bl)
PLANASA	• State sanitation companies dominant • Sustainable pricing with price discrimination • Global viability studies • State participation (up to 5 percent of the tributary revenues) • Adhesion by various municipalities (SP-RS-MG) • Inflation (1979): price restraint and difficulties in obtaining investment returns. • Crisis and questioning of the model between 1985 and 1989.	7.4	1971-1989	3.9	11.3
PRONURB	• Excessive use of FGTS between 1990 and 1991 • Shortage of FGTS resources between 1991 and 1994 • Failure to pay by the governments and state companies • Ongoing debt (Law 8727/93)	2.4	1990-1994	0.8	3.2
PROSANITATION	• FGTS reopening of contracts • Lack of good projects • Low payment capacity • Unsatisfactory operational practices • Launching of the PASS • Federal concessions law (Law 8987/95) • Tariff revenue link to the loans (1996) • 1997: CMN and the contingency for 1998/1999	0.8	1995-1998	0.6	1.4
No main programme	• Discussion of the model to be adopted • Entitlement problem • Important but isolated initiatives	1.9	1999-present	1.8	3.7

The biggest problems in Brazilian sanitation are related to costs and the enormous sum of more than $25 billion that needs to be invested at present to improve the situation. The worst sanitation conditions are in small towns, especially those with fewer than 20,000 inhabitants. It should be noted that Brazil has 4,074 municipalities whose population is smaller than 20,000. This represents about 73.26 percent of the 5,561 municipalities, and according to the IBGE census of 2000, account for a population of 33,618,857 inhabitants, about 20 percent of the total population of country.

In an ideal world, the most neglected percentage of the population, i.e. 33.6 million Brazilians, would be concentrated in a small number of large cities so that the principle of economy of scale can be used to build and/or to increase the sanitation systems. The variable costs would remain the same but the fixed costs would be significantly reduced.

Another serious problem stems from the uncertainty in terms of which level of the government has the necessary institutional authority[7]: state or municipality? The Federal Constitution is ambivalent on this point. It states that it is the responsibility of the municipalities to organise and supply services that are of local interest (art. 30, inc. V). However, it also stipulates that the states may create metropolitan regions, urban sprawls and micro-regions constituted by neighbouring municipalities to operate public functions of common interest (art. 25 § 3°). Sanitation fits in both the situations, thus creating additional complications.

The large sanitation companies in Brazil belong to the states. Their respective revenue bases remain concentrated in the capitals and the surrounding areas, which account for nearly 70 percent of their billing. Therefore, there is a never-ending dispute over entitlements in the large urban sprawls, with the states claiming the concession rights in such areas. Being unhappy with this situation, the municipalities have created an entitlement stalemate, which is at the heart of the problem. Accordingly, Brazil does not have a sanitation policy or law because of this impasse.

In terms of lack of investments, the problem stems from the financial crisis of 1998, which forced the National Monetary Council (CMN) to impose restrictions on public sectors, except the petroleum sector, and, 18 months later, the electricity sector.

Because there was no well-defined law, the sanitation sector became risky for private capital. Private sector became apprehensive and investments dried up, leading to a higher incidence of water contamination. Without public or private sector resources, sanitation services have deteriorated. Sewage treatment services today cover only 25 percent of the urban areas. This percentage drops to about 18% when rural zones around urban areas are taken into account.

This sad situation means that there are still more than 3.5 million urban homes without running water. Although this is not a major national problem in relative terms, the situation is unacceptable and inhuman. The number of houses outside the sewerage network in Brazil is also high, more than 15 million.

---

[7] In Brazilian technical literature, this is referred to as "entitlement."

Finally, it should be emphasised that when this supply and collection problem is solved, the immediate consequence will be an increase in sewage treatment liability, thus further aggravating the problem.

It should also be noted that the institutional water quality problem in Brazil is due to the sanitation sector. This is why the possibility of sanitation being temporarily incorporated into the water sector to solve the problem that affects both the sectors, is being considered. It cannot operate the other way, because institutionally, sanitation sector simply does not exist in Brazil.

6.3 Institutional Arrangements for Water Quality Management

Administration for environmental conservation began in Brazil in 1972, shortly after the United Nations Conference on the Human Environment was held in Stockholm. In 1973, Bahia state created an institution to study the issue. This institution, known today as the Environmental Resources Centre (CRA), has provided a large number of services to the state and many of them have contributed to the improvement in water quality.

In São Paulo, also in 1973, a new company to control water pollution (CETESB) was created to replace the Inter-Municipal Water and Air Pollution Control Commission (CICPAA). Today, CETESB is one of the largest institutions of its type in the world, and is responsible for environmental conservation in a territory that produces about 40 percent of the GNP of Brazil, approximately $300 billion.

So far as the federal government is concerned, in 1973, a Special Environment Secretariat was established. Today, it is known as the National Environment Institute (IBAMA). On January 23, 1986, resolution number 001 of the National Environment Council (CONAMA) required that studies should be carried out on assessing the impacts of several types of industries, including hydroelectric power, on the environment.

Also in 1986, another CONAMA resolution (number 020) defined the criteria for the classification of water bodies. This is an important institutional benchmark in terms of the administration of bulk water quality in the country. A few years later, Brazil started to realise the importance of rational water resources management. Activities in this direction began in 1978 and involved the CEEIBH, which established the benchmarks for water management in Brazil.

In 1988, a new Federal Constitution was promulgated, and this included the establishment of a National Water Resource System. The following year, new state constitutions made provisions for water resources legislations. Eleven states foresaw the need for introducing state legislations on water.

The approval of the São Paulo state water resource law, followed by similar laws in other states, created a favourable climate which allowed the Federal Government to pass Federal Law, number 9433/97. This was possible only after a very long debate, and was finally enacted into law on January 8, 1997. This act established the new National Water Resource Policy, which is implemented jointly by the National Water Resource Council (CNRH), the National Water Agency (ANA) and the states belonging to the federation.

The sequence of events referred to above created several grey areas, of which water quality is one. The responsibilities for water bodies of the country has been divided between the agency or entity that handles environmental management and the agency or entity that deals with water resource management. Although conflicts of interest have been few, the fact remains that there are no clear division of responsibilities between, for instance, ANA and IBAMA at the federal level, between CETESB and DAEE in the state of São Paulo, between FEEMA and SERLA in Rio de Janeiro, and between FEAM and IGAM in Minas Gerais. In practically all the other states, the problem is somewhat similar.

It should be noted that the institutions dealing with water resources management have to grant water use rights, while the entities in charge of environmental management issue environmental licences for businesses. Since the water licence is only legal once the environmental licence has been issued, and vice versa, problems do arise. The solution could be to issue a temporary grant, a legal measure provided for by Federal Law number 9984/2000, until the environmental license can be issued. The states are working on a somewhat similar arrangement. However, it is clear that a single unified process would greatly facilitate the activities of the water resources stakeholder. This is an important point that should be reflected upon in improving the institutional arrangements for water quality management in Brazil.

Another point of concern of an institutional nature is the classification of water bodies by predominant usage. The classification is an important tool for water resources management since it attempts to classify water of inferior quality, which could be used for less demanding purposes.

It is important to emphasise that, so far as the classification tool is concerned, it so should be implemented through agreements with the users of any watersheds, through its committee in terms of objectives that should be satisfied over different time periods. Only under these conditions can the classification be considered to be a useful tool for rational management decision-making.

An example is the work done by the Environmental Resource Centre (CRA), the entity in charge of environmental management of Bahia state, which has monitored and classified the tributaries of the São Francisco River. In spite of the good quality of the results of this study, it has been of little practical use. The main reason is because there are no basin committees in the São Francisco region, there was little participation by the civil society. At present, no one is using the information of the CRA study, irrespective of its quality.

It is important to realise the potential of the basin committees when developing plans on water quality management. The committees should be involved in the decision-making process from the very beginning of the management process.

6.4 PRODES and Water Management

The National Water Agency (ANA) conceived the Catchment Restoration Programme (PRODES)[8]. It was launched in 2001.

It is an imaginative and innovative programme, the type of which is rarely found in public policy projects in Brazil. Its main goals are to restore the quality of water in rivers and lakes and to encourage the use of modern water resource management tools, such as pricing for the use of water and introducing river basin committees and agencies.

Within the context of PRODES, the following undertakings are permissible: (1) new sewage treatment plants, (2) enlargement, retrofitting and upgrading of existing treatment plants, and (3) special sewers which can reduce the present pollution load in water bodies.

PRODES offers financial incentives. The funds, representing 50 percent of capital costs, are deposited in a special account every three months over a 3 to 7 year period, only if proposed project targets agreed to earlier are being met.

PRODES initially intended to invest about $80 million per year over a 10-year period. The objective was to reduce BOD loads by 40 percent during this period. This is equivalent to providing some 37.5 million inhabitants with urban sewage treatment services. This was expected to have a positive effect on public health (Figure 6.1).

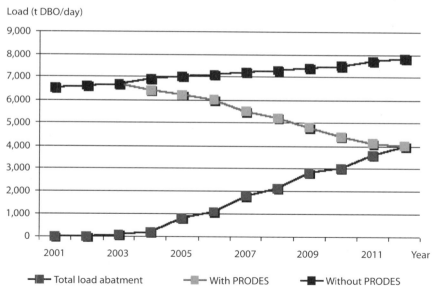

Fig. 6.1. Organic load abatement: PRODES' perspectives (Source: National Water Agency, Brasilia, 2003)

---

[8] PRODES is also popularly known as "sewage buyer," which really means "buyer of treated sewage."

However, significantly reduced allocations of federal funds adversely affected the programme. Instead of improving, the sanitation sector in the country has gone backwards in terms of water quality. In its first year of operation alone, 2001, $18 million were paid out, benefiting about two million inhabitants. Unfortunately, budget cuts meant that this amount was reduced to only $6 million in 2002. By the second half of 2003, approximately $12 million that should have been paid out to the contractors, had not been paid, and the prospects of PRODES getting the necessary funds from the Ministry of Finance were not good.

The objective of this discussion is not on how PRODES works or its merits. Instead, the aim is to initiate a debate on the possibility of the sanitation sector being given, albeit temporarily[9], responsibility for water resource management. The results obtained by PRODES, even though they have been affected by a shortage of funding, require such a national debate.

If this were to happen, the National Water Agency (ANA), which is in charge of implementing the national water policy, could be made responsible for the sanitation policies at the national level. This will delay the need to resolve a federal public administration problem of whether to create a new regulatory agency for environmental sanitation. Although the current debate has not touched on this issue, if the provisions of the Reform of the State Apparatus are to be adhered to, a decision must be taken on how the sanitation issues of the country are to be managed. Otherwise the sanitation problems and, therefore, water quality problems, will worsen with time.

The institutional aspects of water quality in Brazil require an in-depth, objective and independent analysis. The results of such an analysis can significantly improve the overall institutional framework for water quality management.

6.5 Environmental Education

A broad educational programme on environmental management is one good way to improve water quality at the local level. It promotes partnerships and it reduces the costs of management, since the best way to improve water quality is to stop the contamination in the first place. The mobilisation of basin committees has been important in this regard. A programme was initiated in 1995, which soon gained credibility in many states. However, goals could not be met because the basin committees did not function properly because of inadequate operating conditions.

This programme on water, adapted from another one which assures that poor children should go to schools, is based on a rather simple idea: both water and children represent the future of a country, and both are unable to defend themselves from anti-social and industrial activities. Therefore, there has to be a social

---

[9] This temporary situation may last for however long it takes for sanitation service coverage to reach an adequate level and contribute effectively to an improvement in water quality of the country.

mobilisation process in favour of improving water quality and helping the poor children to get a proper education.

Simple actions, such as campaigning for better water quality and calling on people to be vigilant and to report any incidents of water pollution, should be encouraged. In addition, the monitoring of water bodies should be carried out at specified sample points in each basin, and water qualities should be tested regularly.

The partnership established between those who implement the programme and the sanitation companies is vital since the sanitation companies have the greatest interest in getting clean water so as to reduce the treatment costs. The basin committee should be involved, since its role is to ensure the active participation of the citizens and institutions within the basin.

6.6 Conclusions

Despite the recommendations included in the Reform of the Brazilian State Apparatus that was approved in 1996, the state bureaucracy of the country continues to be huge and powerful. To get an idea of its gigantic size, the federal government currently has 36 ministries. Simplification of the bureaucratic machine is urgently needed, not only for the water sector but also for other sectors as well.

A feasible proposal would be that the water resources and sanitation management sectors become one single institution. Since this may have some disadvantages, a multiple-uses principle should be applied and honoured. For example, one of the sectors related to water resources should be responsible for water management decision-making, but that the interests of no single sector should dominate the overall decision-making process. In addition, no one single use should prevail over the others, except in the case of domestic water supply for human consumption. This will require compromises between different sectors.

The proposal to have a single institutional structure should also be helpful to study in greater depth the existing regulatory system for sanitation, both technically and institutionally. It would also encourage the discussion of the entitlement issues, starting with the basin committees, where civic institutions, the water users and municipalities are involved. The debates at the basin committee level should contribute to rational decision-making at different governmental levels. None of these are happening satisfactorily at present.

6.7 Bibliography

Azevedo LG, Simpson L (1995) Brazil: Management of Water Resources. Economic Notes 4, The World Bank, Washington

Barraqué B (1995) Les Politiques de l'Eau en Europe. Éditions La Découverte, Paris

Barth FT (1991) Modelos para Gerenciamento de Recursos Hídricos. Coleção ABRH, vol I, São Paulo

Biswas AK (1996) Water Development and Environment. In: Biswas AK (ed) Water resources, Environmental Planning, Management and Development, McGraw Hill, New York, pp 1-35

Carrera-Fernandez J, Garrido R (1999) Economia dos Recursos Hídricos. Teoria, Metodologia e Estudos de Casos, Salvador

Formiga R, (1998) Les Eaux Brésiliennes vers une Gestion integrée: Une Analyse des Réformes Féderale et de l'Eau de São Paulo. Thèse de Doctorat en Sciences et Techniques de l'Environnement, l'Université Paris XII – Val de Marne, André, Paris

Garrido R (1993) Problèmes des Ressources en Eau au Brésil. Séminaire de Gestion des Ressources en Eau à l'Université Laval, Québec, Canadá

Garrido R (1998) Gestion des resources en eau au Brésil. Conférence Internationale Eau et Development Durable, Paris

Gordilho-Barbosa AM (2004) La Réorganisation des Services Publics d'Eau et d'Assainissement dans l'État de Bahia au Brésil: un enjeu multiniveaux, une ouverture aux acteurs sociaux. Thèse de Doctorat Université Paris XII – Val de Marne, Bernard, Paris

Leme-Machado PA (2002) Direito Ambiental Brasileiro. Atlas, São Paulo

Secretaria de Recursos Hídricos (2002) Política Nacional de Recursos Hídricos. Lei n° 9.433 de 8 de Janeiro de 1997. Ministério Do Meio Ambiente, Brasilia

Secretaria Do Meio Ambiente (1997). Gestão das Águas: Seis Anos de Percurso. Secretaria de Recursos Hídricos, Saneamento E Obras. Governo de São Paulo, São Paulo

Tomanik-Pompeu C (1994) Aspectos Jurídicos da Autorização Administrativa. Anais do X Seminário-Curso do Programa Centro Interamericano de Recursos da Água – CIRA, Salvador

Tomanik-Pompeu C (1997) Aspectos Jurídicos e Normativos da Outorga de Direito de Uso dos recursos Hídricos: A Lei Nacional ante as leis estaduais. Texto apresentado No. I Encontro Técnico sobre Cadastro de Usuários e Outorga de Direito de Uso dos Recursos Hídricos, Fortaleza

Water Quality Management in Ceará, Brazil

José Nilson B. Campos and Francisco de Assis de Souza-Filho

7.1 Introduction

Because water resources management occurs within the overall context of human values and physical realities, each society develops its own systems and goals. Perception of societies regarding natural resources, reflect biophysical realities, cultural values, historical experiences and political realities (Perry and Vanderklein 1996).

The evolution of the water quality management system of Ceará state represents a good example. A first stage included the water system, which was developed in the context of scarcity and recurrent droughts: the sole purpose was search for water. Political and technical actions concentrated on drilling wells and building reservoirs to store water to combat the droughts. During this hydraulic phase, as it became known later, many reservoirs were built in the semi-arid regions of Brazil.

In the second phase, government action was aimed at using the hydraulic potential of the reservoir. Major irrigation projects were undertaken, several cities had water systems built or improved, and industrial development took place. Nevertheless, all these measures were taken with little, if any, concern for the treatment and proper disposal of wastewater. The consequence was a decline in the water quality of many reservoirs, lakes and aquifers. In some cases, mainly in small reservoirs, the concentration of pollution made the water unusable for human consumption. Poor people, who were forced to use this contaminated water, suffered from serious health problems.

During the third and current phase, there is some concern to manage water quality properly, and society is trying to build an efficient management system. This phase began in 1992, with the completion of the State Water Resources Master Plan (SRH 1992). The plan provided the basis for an integrated system of water resources management and modified the institutional framework for water management in the state. Although the main part of the plan addressed quantity and allocation, it also introduced the concept of integrated management of quality and quantity. Additional improvements are necessary to build an effective water quality management system.

This chapter presents an assessment of the evolution of water quality management in the Ceará state, describing the physical conditions, cultural aspects, evolution of the legal and institutional frameworks, and the challenge of developing a truly integrated water management system.

7.2 Physical Context

Ceará is located in northern Brazil (Figure 7.1). The state is characterised by a severe climate and a hydrologic regime that has been historically an obstacle to human settlements and development. This section describes the physical realities of the region in terms of climate, biota and hydrologic and cultural aspects.

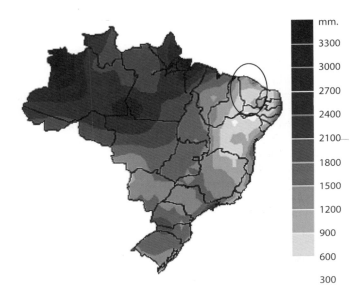

Fig. 7.1. Map of Brazil highlighting Ceará, and showing average annual rainfall in mm

7.2.1 Climate

The average annual rainfall ranges from 1,400 mm on the coast and mountains to about 500mm inland (Sertão Central). More than 90 percent of annual precipitations is concentrated in six months (December-June) and more than 75 percent occurs in four months (February-May). Evaporation in some places reaches more than 2,500 mm. These conditions are combined with a crystalline soil that results in the formation of a network of intermittent rivers. In a normal year, the water balance is positive only during three to four rainy months. This wet period is used by *sertanejos*, as the people of the region are known, to cultivate subsistence crops, mainly beans and maize.

7.2.2 Hydrologic Regime

Most of the rivers of the north of the Brazilian Northeast stay dry for six to nine months each year, or even more than a year when severe droughts occur. The histogram of monthly means discharges at Jaguaribe River (Figure 7.2) and its daily discharges in the drought year of 1958 (Figure 7.3) illustrate the hydrologic characteristics of the region.

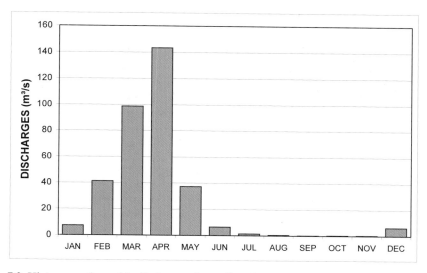

Fig. 7.2. Histogram of monthly discharges, Jaguaribe River, Iguatu, Ceará, 1912-1993

Describing the 1903 drought at Rio Grande do Norte, Guerra (1981: 131) stated that:

"Rivers, such as the Mossoró, have been dry for 24 months. Since the next six months are summer, it will continue upto December like this, and it is frightening to think that a 360 km river, more or less the length of the Thames, can stay dry for 30 consecutives months. .... We are in July now, and already several farms are without water. Many *sertanejos* have to travel 3, 6 and even 12 km in order to find drinking water."

In addition, the annual variations of river discharges are among the highest in the world. The coefficient of variation of annual inflows ranges from 0.6 to 18. In temperate regions with perennial rivers, these values are usually around 0.3-0.4. This high variability implies the need for large reservoirs to provide some reliability in the provision of water supply.

The state has problems of water quality associated with the natural salinisation of some reservoirs. In Ceará, soil is the main cause of the salinisation. Reservoirs, whose catchment areas contain soils of the type *Planosolos Sólodicos* and *Solonetz Solodizado,* contribute high concentrations of salts to their waters. As a result, surface water sources have to be carefully administered in order to maintain salt concentrations at acceptable levels. The problem of salinisation is more acute in water from underground sources that are in contact with the crystalline rocks.

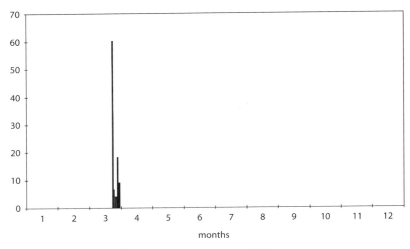

months

Fig. 7.3. Daily discharges (m^3/s) at Jaguaribe River in 1958

7.2.3 Biota

Phytoplankton is the basis of the food chain, and zooplankton is the subsequent link. Management of water quality should promote the control of algae and pathogenic organisms. The presence of algae, and the manner of their occurrence, can serve as indicators of the trophic state of a lake, besides pointing to possible presence of nutrients.

Excessive presence of algae in the water can represent problems, though the toxins produced are usually harmless to both humans and animals. They can liberate composts that affect the flavour, colour and odour of water, besides contributing to operational problems of the treatment stations, where, for example, they could block filters. In general, water, with an excessive presence of algae, tends to cause damage to human health because of the increased likelihood of the formation of harmful by-products of chlorination.

The most commonly identified algae species found in reservoirs include some that cause problems related to the taste and odour of the water, or the obstruction of treatment filters. The following types have been identified in the state:

1. *Anabaena* and *Aphanizomenon* alter the taste of the water, form granular obstructions and are toxic. Taken out of the water, they smell like radishes.
2. *Oscillatoria, Staurastrum* and *Synedra* alter the taste of the water, cause obstructions and smell like grass.
3. *Cyclotella* and *Trachelomonas* commonly cause obstructions.

4. *Dictyosphaerium* is highly aromatic, with a radish-like odour.
5. *Ulothrix* smells like grass.
6. *Mycrocystis* is poisonous and belongs to the group of blue algae. Some stumps or strains of these algae produce a relatively potent endotoxin that can produce death in one hour when inoculated in the proportion of 0.5 milligrams by kilo of weight of the animal.

Zooplanktons found in the reservoirs are:

1. *Arcella*, a protozoan of the *Sarcodina* class of amoebas, that can be parasitical. The pathogens it causes are known as *dysentery amoebae*.
2. *Difflugia* is also a protozoan of the *Sarcodina* class, and is sometimes parasitical.
3. *Rotifera* are common in calm freshwater with abundant aquatic vegetation. They help to maintain the water clean by feeding on organic debris and other organisms. They also serve as food for worms and small crustaceans.
4. *Bdelloidea, Filinia, Karatella* and *Lacame* belong to rotifer species.
5. *Copepoda* are perhaps the most important crustaceans because they are herbivore, feeding on phytoplankton and, therefore, constituting the base of most food chains. However, they also include a great number of fish parasites that make the link to other animals.
6. *Cyclopoida* is a *copepoda*-type sponge.
7. *Cladocera and Calanoida* are crustaceous filters.

7.2.4 Cultural Aspects

Climatic and hydrologic uncertainties and adversities have forged *sertanejo* culture. Guerra (1981) described the feelings of the *sertanejos* who live off rainfed agriculture: "The *sertanejo* is always frightened and afflicted with fear of drought. In October, the anxiety and ill-feelings begin. Will it rain? Will drought come? Should we till the land? Will there be any point in investing money?"

The ever-present threat of drought created among *sertanejos* a tradition of storing water during the rainy season in order to use it sparingly during the dry periods. No farm in the Brazilian Northeast can be sustainable unless a permanent water source is nearby. According to Guerra (1981): "In the sertão, it is better to leave the family a good water hole than a rich and beautiful palace." Scientists and engineers in the region are nearly unanimous on the need for reservoirs as an efficient means for mitigating the impacts of droughts. Some even say that reservoirs should be as large as possible in order to control all the water in such a way that not a drop reaches the ocean (Guerra 1981: 42).

7.3 Political Realities

Politics in the Brazilian Northeast, particularly in Ceará, have long been drought-driven, and still are to a certain extent. Whenever a drought occurs, politicians come to the state to announce new measures and public works, mainly construction of reservoirs, wells and irrigation projects. Few of their promises, however, become reality.

Sometimes the promises are dramatic. In 1877, after a particularly severe drought, whose impact shocked the nation, Brazilian Emperor Pedro II, said: "I will sell the last jewel in my crown rather than let anyone from the Northeast die of hunger." Action, however, takes longer and more difficult. For example, the Orós Dam in the Jaguaribe River was completed in 1965, but the valve to release the stored water was not ready to work for another 15 years (Guerra 1981: 142).

Measures to improve the management of water quality are not politically efficient (they produce little return in the ballot box), so politicians tend to shun them. This is an important reason for the delay in building an efficient water quality management system.

7.4 Institutional and Legal Frameworks

The Brazilian Water Code, promulgated in 1934, was the first landmark in the legal framework of a national water management system. Nevertheless, the Code was formulated on the basis of the climatic and hydrological realities of the Brazilian South and Southeast. Article 3 establishes that the rivers should be perennial in order to consider their water to be in the public domain. Hence, according to this article, none of the rivers in Ceará would be in the public domain, since all of them become dry every year. Nevertheless, the Code recognises the peculiarities of hydrology in semi-arid climates, and, in article 5, it grants public access for common uses to all water in regions periodically devastated by droughts. Taking into consideration specific legislations on drought mitigation policies, it also considers this water to be in the public domain.

The water policy of the federal government in the Northeast took its lead from a legal framework that was based on a policy of combating drought. Article 177 of the 1934 Federal Constitution established that "Defence against the effects of drought in the states of the North will follow a systematic plan that will be permanent and under federal control. The national government will spend on public works and attendant services on an amount not less than 4 percent of fiscal income that has not already been assigned."

Law 175 of 1936 considered as public works and attendant services as:

1. Regularisation and diversion from rivers of water for irrigation and other uses, including canals, dams, mechanical elevation, drainage of irrigated areas, and any complementary or related services;
2. Drilling of wells and construction of galleries to capture water;

3. Fish farming in rivers, lakes and reservoirs;
4. Establishment of forests;
5. Studies and systematisation of irrigation methods in accordance with farmers' capacity;
6. Construction and maintenance of roads required for the execution and efficient use of the works and services;
7. Systematic collection of data on geology, hydrology and meteorology in the area of reference; and
8. Provision and organisation of statistics based on data collected (Campos 2003).

At that time, DNOCS (National Department of Public Works Against Drought) was the federal institution that was in charge of building and maintaining the public works in question. Water quality management had no place in the legal framework. The main concerns were construction and maintenance of water works, regulation of use of water in accordance with economic activities, and improvements in knowledge of the geophysical characteristics of the region.

Concern about water quality management emerged in the context of Resolution 20 of the National Council on the Environment (CONAMA), which can be considered to be a landmark in establishing desirable standards for water quality. The resolution includes two basic concepts: the classification of water bodies and then matching water bodies with the appropriate relevant classifications.

The classification relates to the establishment of water quality standards in accordance with use. Nine classes of freshwater are identified.

The application of Resolution 20 of CONAMA to rivers that flow only intermittently involves several conceptual difficulties. The main problem is the definition of a reference flow to calculate the loads in the river. The definition presented in the Resolution produces a result of zero for intermittent rivers, and needs to be revised to provide norms very specifically for the semi-arid Northeast.

7.4.1 Ceará State Law

The first Ceará state law on the preservation and control of water resources was promulgated in 1977 (Law 10.148, December 2, 1977). In this law, article 6 prohibits the release of pollutants into the water. A pollutant is defined as any kind of matter or energy that causes contamination. However, article 8 establishes that "the installation, construction or expansion, as well as the operation of sources of pollution, requires prior authorisation of the competent agency of the state."

Articles 6 and 8 appear to be contradictory, since the first forbids the release of pollutants in water bodies, reflecting fears that surface reservoirs may be seriously contaminated by receiving only polluted water, while the second reflects the certainty that the release of some pollution is inevitable. This anomaly continues to persist. Four years later, another law on preservation and control of water resources was promulgated (Law 14.535, July 2, 1981). It retained the same apparent contradiction.

Later, the 1992 state water law (Law 11.996, July 24, 1992) introduced the concepts of permits and charging for the release of wastewater into a water body. Article 7 established that charges for the dilution, transport and assimilation of wastewater must take into consideration the class of use of the receiving water body, the level of regulation by reservoirs, and the amount of pollutant released, including expected variations. The physical, chemical and organic parameters of effluents are defined. While the text covers the problem of release of pollutants in broad terms, it can mean very different things to different people in quantitative terms.

7.5 Building a New Water Quality Management System

According to Grigg (1996), a water quality management system includes the control of drinking water, in-stream and groundwater quality, and effluents from wastewater plants and nonpoint source discharge points. In order to construct a proper water quality management system, it is necessary to establish criteria and standards for the monitoring and assessment of water quality, treatment plants, and water quality databases and quality modelling. With these requirements in mind, the issue that needs to be analysed is whether the Ceará water quality management system is working properly.

The objectives of the institutional water resources and environmental systems are different and potentially conflicting. The perspective of the former is water is essential for satisfying human needs, while the aim of the environmental system is the promotion of nature as an end by itself. Despite the potential conflict, the two systems can complement each other on specific issues.

The system of water resources also differs from the system of basic sanitation. Basic sanitation is a sectoral use, as are irrigation and energy. Environmental sanitation policies and measures taken to enact them should be defined within water policies and vice-versa.

The administration and management of water quality needs to be developed within six dimensions: institutional outline of standards, standards system itself, planning, licensing, monitoring and control system, and finance.

7.5.1 Institutional Framework

The Ceará system of administration of water resources is composed by i) a system of representative bodies (the State Council of Water Resources and Basin Committees; and, ii) an executive system of the state policy coordinated by the Secretary of Water Resources, who oversees the functioning of the Company for the Administration of Water Resources (COGERH) as administrative organ and supervisor. Part of this system is the Ceará Foundation for Meteorology and Water Resources, which has a role in monitoring and in conducting basic hydrometeorological studies in the areas of water resources and the environment.

The state system incorporates the participation of the main federal organs involved in the area of water resources, especially DNOCS. It also tries to incorporate environmental issues through the Office of the State Superintendent for the Environment, which executes policies dictated by the State Secretary of Public Works and the Environment. The State Environment Council has to approve all major projects.

Integration of the water management and environmental systems should lead to:

1. Establishment of a basic network for monitoring water quality;
2. Definition by the basin committees of goals for water quality;
3. Clear definition of the action to be taken by each system;
4. Assumption of responsibility for the coordination of the financial and operational efforts that are needed to meet the established goals; and
5. Establishment of a specialized technical team for planning and conducting basic studies.

7.5.2 Regulations

The national norms for water quality, while adequate for humid areas, falls short of what is needed for semi-arid regions such as Ceará. Progress on the establishment of adequate norms for semi-arid regions will, however, depend on the acquisition of more knowledge on the hydrodynamic, physio-chemical and biological characteristics of the lakes and the intermittent rivers of the state.

Although the formulation of norms is difficult, measures to ensure their implementation (whether through coercion or convincing) represent even more of a challenge. Participation by the public in basin committees is helpful in this context, as are forums that can create a consensus on the types of actions that could be taken. Even so, a structure has to be created that guarantees control, through coercion, if and when necessary.

7.5.3 Planning

Water quality planning should seek strong integration with urban and land use policies, and to provide guidelines on urban water supply, sewage, drainage and disposal of solid wastes. Land use in both rural and urban context should be debated together with the definition of water quality goals. It is, indeed, impossible to achieve effective water quality policies in large human settlements, without appropriate discussion of land use.

Brazilian legislation includes an instrument for this purpose, though it has not been properly used, especially in semi-arid regions. It is the definition of watercourses. In discussing such definitions, we should define the type of river that we want to have in the long term and the steps that should be taken to reach that objective.

7.5.4 Licensing and Regulation

Planning is put into practice by the licensing process. Where the licensing process is ineffectual, planning is a waste of time. Regulation is a tactical mechanism through which planning goals are put into effect by controlling the behaviour of polluters.

7.5.5 Monitoring and Control

What guarantees the operation of licensing, regulation, and hence, planning, is a system of monitoring and control. This system provides feedback on planning by generating indicators on system performance that can point to the need for adjustments.

7.5.6 Finance

What can be achieved in terms of water quality for several uses is strongly conditioned by the question: who pays the bill for maintaining the water quality of a specific river? The fiscal crises of the 1980s and 1990s have sharply reduced the capacity of the government to fund projects. Therefore, charges for the use and the pollution of water should be part of a mechanism that, besides promoting rational use, can contribute to the generation of funds that can be used for planning and implementing specific initiatives.

7.5.7 Comments

In Ceará, water quantity has been monitored and evaluated only since 1993. Six years later, some efforts were made to improve the methodology used to monitor water quality, and only in the last two years a water quality monitoring network has been set up (COGERH 2002). However, only the hydrographic system used to supply the metropolitan region of Fortaleza has good coverage of physical, chemical and biological analysis.

Analysis and modelling of water quality from reservoirs and underground sources are still lacking. The problems of seasonality and of the long periods during which rivers are dry have not been properly evaluated.

More effective measures that have still to be taken for the protection of water supply catchments include:

- Regulatory measures (zoning);
- Regulation for wastewater discharges;
- Regulation for solid wastes management; and
- Environmental education.

A number of rivers in Ceará have now been made artificially perennial because they receive wastewater from cities. The old concept of not allowing one single drop of water flow into the sea is no longer valid. More studies are needed to measure flow and schedule flows and discharges.

7.6 Conclusions

The water management system in Ceará state emerged within a context of scarcity and droughts. On the quantitative side, the model was well designed and has shown significant improvements since 1992. Regarding quality, there has been an attempt to establish a monitoring network only during the last two years. An enormous amount of work remains to be done, including:

- A more effective use of the data that is being collected;
- An accurate assessment of water quality in the reservoirs;
- Models of water quality in reservoirs in order to better evaluate the impacts of pollutants;
- Determination of environmental flows in order to dilute and assimilate pollutants properly, and for ecosystem needs;
- Investment for water treatment, wastewater disposal, solid wastes management and regulations; and
- Despite the advances of recent years, water quality management is not working properly in Ceará. Significant improvements, studies and investments are necessary. To build an ideal water quality management system within the context of financial restriction is a big challenge for all those involved.

7.7 References

COGERH (2002) Rede de Monitoramento de Água Operada pela COGERH, Fortaleza, Ce Brazil

Campos LR (2003) A Proteção dos Recursos Hídricos pelo Estado. Imprensa Oficial Graciliano Ramos, Maceió, Al, Brazil

Guerra PB (1981) A Civilização das Secas. Departamento Nacional e Obras Contra as Secas, Fortaleza, Ce, Brazil

Grigg N (1996) Water Resources Management, Principles, Regulations and Cases. The McGraw Hill Company, New York

Perry J, Vanderklein E (1996) Water Quality: Management of A Natural Resource. Blackwell Science, Cambridge, Massachusetts

SRH (1992) Plano Estadual de Recursos Hídricos. Secretaria dos Recursos Hídricos do Ceará, Fortaleza, Ce Brazil

Water Quality Management in Mexico

Felipe I. Arreguín-Cortés and Enrique Mejía-Maravilla

8.1 Introduction

Average annual rainfall over Mexico's two million square kilometres of territory is 771 millimetres, equivalent to 1,511 km^3. Distribution, however, is uneven. In 42 percent of the territory, mainly in the north, annual average rainfall is less than 500 mm and as little as 50 mm in some places, such as in the vicinity of the Colorado River. On the other hand, in 7 percent of the territory, the annual average exceeds 2,000 mm, with some regions registering more than 5,000 mm. Nearly all the rainfall occurs in over just four months, 80 percent of it during the summer. Approximately 27 percent of the total rainfall is transformed into runoff, of which 399 km^3 a year on average flows through the nation's 320 river basins. Here too, distribution is highly uneven, since 50 percent of the runoff is generated in the south-east, which accounts for 20 percent of the nation's territory, while just 4 percent occurs in the north, which accounts for 30 percent.

Mexico's lakes and lagoons have a total capacity of 14 billion m^3, while dams store a further 150 billion m^3. These joint capacities cover about 47 percent of the total annual runoff. Some 67 billion m^3 of rainfall filters through to form the renewable resource of the aquifers, while aquifers located below irrigated areas receive an artificial recharge of 15 billion m^3. One-off capacity of the aquifers is rated at 110 billion m^3.

The amount of water available per capita is extremely variable on a regional basis. Some regions show values of between 211 and 1,478 m^3 per year per capita, while others fluctuate between 14,445 and 33,285 m^3. The national average is 4,534 m^3.

For water management purposes, Mexico has been divided into 13 regions based upon the existing hydrologic divisions, as shown in Table 8.1. In terms of the natural availability of water, regions X and XI register values of 10,604 and 24,674 m^3 per capita per year, respectively. In contrast, region XIII barely has 182 m3 per year, while region VII is 1,729 m^3 (Table 8.1). Unfortunately, however, uneven distribution and seasonal factors are not the only restrictions on water use. In recent years, pollution has become the main constraint.

8.2 Water Quality

Water quality in surface water, coastal zones and aquifers is monitored periodically through the National Water Quality Monitoring Network. This network has 362 permanent stations, 205 of them located in surface water bodies,

44 in coastal zones and 113 in aquifers. The secondary network has 276 mobile stations, 231 of them located in surface water bodies, 17 in coastal zones, and 28 in aquifers. In addition, a reference network of 104 groundwater stations keeps track of the evolution of quality in aquifers.

Table 8.1. Natural water availability per region.

	Administrative region	Average natural availability hm^3	Average natural availability per capita (2003) m^3/inhab	Total natural runnoff^a hm^3	Average recharge in aquifers hm^3
I	Península Baja California	4,423	1,336	3,012	1,411
II	Noroeste	8,214	3,236	5,459	2,755
III	Pacífico Norte	24,741	6,035	22,160 ^{b}	2,581
IV	Balsas	28,909	2,713	24,944	3,965
V	Pacífico Sur	33,177	7,963	31,468	1,709
VI	Río Bravo	13,718	1,324	8,499	5,219
VII	Cuencas Centrales, Norte	6,836	1,729	4,729	2,107
VIII	Lerma-Santiago-Pacífico	38,264	1,729	30,954^b	7,310
IX	Golfo Norte	23,347	4,685	22,070	1,277
X	Golfo Centro	102,546	10,604	98,930	3,616
XI	Frontera Sur	157,999	24,674	139,578	18,421
XII	Península de Yucatán	29,063	8,178	3,747 ^{b}	25,316
XIII	Aguas del Valle de México y Sistema Cutzamala	3,803	182	1,996 ^{c}	1,807
National level		475,040	4,534	397,546	77,494

Source: CNA, 2004.
Notes: Runoff values are annual average values according to historical records.
[a] Including water imports and excluding water exports.
[b] Preliminary information, studies for these regions have not yet been concluded.
[c] Wastewater from Mexico City is included.

8.2.1 Classification Studies

A number of factors can modify water quality in river basins. From the viewpoint of sustainable development, current challenge is about achieving a balance between productive activities and water quality for several water uses. Classification studies for aquatic systems are water quality assessments that allow the authority to set quality goals in short, medium and long terms, based on the capacity of water bodies to dilute and assimilate the pollutant loads that are discharged into them (UNESCO-WHO 1978). These studies are planning tools that help the water authority to develop corrective actions intended to achieve the cleaning of water bodies as well as the prevention of polluting discharges.

With this perspective, 18 studies are being performed, or are being updated, in order to estimate the capacity to dilute and assimilate the pollutant loads of different water bodies. This is done in accordance with the National Water Law (SEMARNAT 2004).

8.2.2 Biological Indicators of Water Quality

In order to enrich water quality information with sediment and biotic data, some biological approaches that work with benthic organisms and ecotoxicological tests are being used.

The most common way to monitor water quality in an aquatic system with a biological approach is through the community of macro invertebrates, benthic which include a number of organisms such as shells, clams, crabs, larvae, nymphs, adult aquatic insects, leeches, and aquatic worms, all of which become biological indicators for water quality. They serve as references to a sustainable management of water resources and provide information that allow the authority to set regulatory guidelines on specific issues affecting water quality.

A National Programme for Biological Monitoring was launched in 2003, and the initial results from monitoring sites located in some selected rivers are presented in Figure 8.1. Comparative Sequential Index has been defined as the biological indicator that can be used.

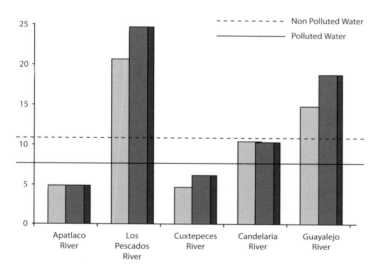

Fig. 8.1. Results of National Programme for Biological Monitoring, 2003. ISC Values: <8 Polluted water; >8 <12 Partially polluted water; >12 Non-polluted water.

8.2.3 Pollution of Water Bodies

Table 8.2 presents the most polluted water bodies in Mexico, according to the results of an evaluation based on the Water Quality Index (WQI) of 2002.

Table 8.2. Highly polluted water bodies

Administrative region	Basin	Highly polluted rivers
I Península Baja California	Tijuana – Maneadero Colorado	Tijuana and Tecate Nuevo
III Pacífico Norte	San Pedro	Arroyo Acequia Grande; Durango; Tunal Súchil
IV Balsas	Atoyac	Zahuapan; Atoyac and Alseseca
VI Río Bravo	Laguna de Bustillos y de Los Mexicanos	Laguna de Bustillos
VIII Lerma Santiago Pacífico	Lerma –Toluca Lerma – Salamanca Laja Santiago – Guadalajara Verde – Grande	Lerma Lerma and Turbio Querétaro Santiago Lagos
IX Golfo Norte	Moctezuma	San Juan and Tulancingo
X Golfo Centro	Papaloapan	Blanco
XIII Aguas del Valle de México y Sistema Cutzamala	Moctezuma	de los Remedios; Churubusco; San Buenaventura; de la compañía; Teotihuacan; Presa Heñido y de las Avenidas

Source: CNA, 2003a.

Wastewater discharges from urban areas are presented in Table 8.3.

Table 8.3. Wastewater discharges and pollutant loads in urban areas per year

Items	Annual Volume
Total wastewater produced	7.95 Km^3 (252 m^3/s)
Wastewater collected in sewer systems	6.40 Km^3 (203 m^3/s)
Total pollutant load produced	2.17 million tons of BOD
Pollutant load collected in sewer systems	1.67 million tons of BOD
Pollutant load removed in wastewater treatment plants	0.38 million tons of BOD

Source: CNA, 2004.

Industrial discharges are presented in Table 8.4, and industrial sectors with the major pollutant loads to watercourses are presented in Table 8.5.

8.2.4 Municipal Wastewater

In December 2002, municipal sewage systems in Mexico collected a total wastewater flow of 203 m^3/s, of which only 29 percent (58.9 m^3/s) was treated. In municipal wastewater treatment facilities, about 23 percent of the organic load, in terms of BOD$_5$, was removed (Figure 8.2 and Table 8.6).

Table 8.4. Industrial discharges and pollutants loads per year

Items	Annual Volume
Total wastewater produced	5.39 Km^3 (171 m^3/s)
Pollutant load produced	6.30 million tons of BOD
Pollutant load removed by wastewater treatment plants	1.10 million tons of BOD

Source: CNA, 2004.

Table 8.5. Industrial wastewater volumes by sector

Type of Industry	Volume (m^3/s)	Organic matters (10^3 tons/ year)
Aquaculture	67.6	7
Sugar	45.9	1,750
Petroleum	11.4	1,186
Services	10.3	183
Chemical	6.9	406
Pulp and paper	5.5	108
Agriculture and cattle raising	3.2	1,063
Food	3.0	193
Beer and malt	1.6	272
Mining	0.8	56
Textile	0.7	14
Distilling and wineries	0.4	230
Coffee	0.3	32
Leather processing	0.1	9
Other branches	12.9	795

Source: CNA, 2004.

Table 8.6. Municipal wastewater treatment plants by administrative region (December 2002)

	Administrative region	No. of operating plants	Design flow (L/s)	Treated flow (L/s)
I	Península Baja California	30	5,519.1	4,655.5
II	Noroeste	69	3,377.4	2,407.6
III	Pacífico Norte	99	5,837.6	4,226.2
IV	Balsas	83	5,681.2	4,181.0
V	Pacífico Sur	52	3,420.4	2,135.6
VI	Río Bravo	118	20,820.2	15,548.9

VII	Cuencas Centrales del Norte	54	1,739.7	1,168.6
VIII	Lerma-Santiago-Pacífico	313	15,902.6	11,650.8
IX	Golfo Norte	47	883.0	680.2
X	Golfo Centro	74	2,838.0	995.0
XI	Frontera Sur	40	1,162.8	900.0
XII	Península de Yucatán	36	1,812.5	1,206.8
XIII	Aguas del Valle de México y Sistema Cutzamala	62	10,740.6	6,392.2
	Total	1,077	79,735.0	56,148.5

Source: CNA, 2003b.

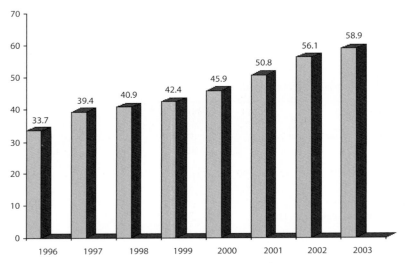

Fig. 8.2. Municipal wastewater with treatment, 1996-2002 (m³/s), CNA, 2003b

8.2.5 Industrial Wastewater

In December 2002, Mexican industries generated a total wastewater flow of 171 m³/s, of which 15.3 percent (26.2 m³/s) was treated (Figure 8.3). In the industrial wastewater treatment facilities, about 17 percent of the organic load, in terms of BOD_5, was removed (Figure 8.3 and Table 8.7).

8.3 Groundwater Pollution

Groundwater is contaminated by human activities. This development is increasingly constraining the development of a number of river basins.

Depending on agricultural practices and subsurface characteristics, aquifers under agricultural lands have been contaminated in varying degrees by soil washing and improper use of pesticides and fertilisers. Aquifers located below urban and industrial zones have become contaminated because of wastewater infiltration, chemical spills and leaching of solid wastes.

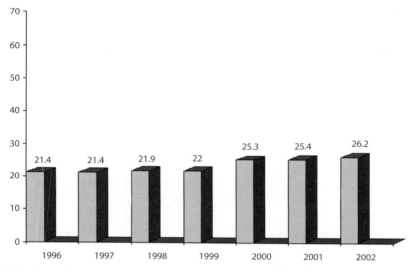

Fig. 8.3. Treated industrial wastewater in 1996-2002 (m^3/s), CNA, 2003c

Table 8.7. Industrial wastewater treatment plants by administrative region (December 2002)

	Administrative region	No. of existing plants	No. of operating plants	Design Flow (L/s)	Non municipal wastewater discharge (L/s)	Treated Flow (L/s)
I	Península Baja California	191	164	1,189.8	1,879	1,102.1
II	Noroeste	20	18	303.6	22,304	83.6
III	Pacífico Norte	30	26	685.6	42,741	468.7
IV	Balsas	226	206	2,933.8	14,778	2,058.0
V	Pacífico Sur	15	14	228.8	4,390	225.0
VI	Río Bravo	99	97	5,008.2	4,217	3,293.1
VII	Cuencas Centrales del Norte	92	92	1,201.4	2,811	824.0
VIII	Lerma-Santiago-Pacífico	348	344	3,905.7	11,059	2,730.8
IX	Golfo Norte	62	61	2,080.3	14,272	1,391.0
X	Golfo Centro	190	186	13,628.7	36,692	11,698.7

XI	Frontera Sur	79	77	1,116.6	7,409	1,070.5
XII	Península de Yucatán	120	108	217.0	2 427	119.9
XIII	Aguas del Valle de México y Sistema Cutzamala	55	55	1,804.1	6,321	1,166.2
	Total	1,527	1,448	34,303.6	171,300	26,231.6

Source: CNA, 2003c.

Biological pollution is common in rural areas because of a lack of basic sanitary facilities or poor building standards. It is common to find wells that have no sanitary protection even though they supply water for human consumption. In addition, such wells are often located near to primitive septic tanks.

Currently, 80 percent of 653 aquifers in a nationwide classification have total dissolved solids (TDS) concentrations of less than 1,000 mg/l, which means availability of good quality water for drinking purposes. However, water is pumped and used with TDS concentrations over this level in some areas where water is scarce.

Severe public health issues, real or potential, arise in several basins, mainly in the north and centre of the country, due to the presence in the rocks having natural chemical elements such as arsenic, fluoride, iron and manganese that are dissolved in groundwater. Consequently, ambient concentrations exceed quality standards.

In coastal zones, as a result of improper water management, around 18 aquifers are facing seawater encroachment, most critically in the north-west (the Baja California peninsula and Sonora). In these aquifers, seawater has progressively moved into the mainland, impairing the use of many wells. This meant that the salinity of groundwater has increased to concentrations that make it unsuitable for most common uses and this trend is probably irreversible.

Another relevant issue that stems from human activities is the pollution of aquifers due to oil and gas spills. More than 50 sites have been registered where some studies have been conducted in order to assess the problems and to propose corrective measures. Even though the impact of those spills is local and temporary, the occurrence of these events in aquifers poses a serious risk for the big cities which obtain their mater water from such sources.

8.3.1 Prevention and Control of Aquifer Pollution

The damage to groundwater quality caused by agriculture development, and the risk posed to human health are being evaluated. Pollution implications due to agriculture practices are being assessed. This includes estimation of pollutant loads, soil attenuation, diffusion of pollutants in the aquifer, impacts on water sources and health risks. Preventive and corrective measures are being considered.

Important progress has been made in setting technical bases for: i) guiding the design and operation of groundwater quality monitoring networks, ii) determining aquifer vulnerability to pollution by specific criteria, iii) assessing pollutant diffusion in subsurface and, iv) defining the most suitable methods for the cleansing of aquifers under different conditions. This is one of the issues to which most attention must be paid in the coming years since in many aquifers water quality problems have become more important than quantity ones, representing a constraint for future human development.

The National Water Law (SEMARNAT 2004) provides a means to help to reduce the damage to groundwater quality by introducing the legal concept of responsibility for the environmental damage. The Law clearly sets that those responsible for pollution regarded by the authority as irreversible must provide full economic compensation for the resulting damages.

8.4 Regulatory Framework

8.4.1 National Water Programme, 2001-2006

The programme highlights the mechanisms and instruments that promote and foster water conservation and sustainable development for every river basin by improving existing water management practices and processes. The goals related to water sanitation set by the programme are showed in Table 8.8.

Table 8.8. Goals for the 2001-2006

Indicators	Goals (%)					
	2001	2002	2003	2004	2005	2006
Population served with sewer system	76	77	77	77	78	78
Treated wastewater / collected wastewater	23	28	31	36	60	65

Source: CNA/SEMARNAT, 2001.

These goals are being achieved through an increase in the income of water utilities as a result of increased efficiency gains and higher water tariffs.

The programme encourages greater participation by private companies in order that additional investments in the water and sanitation sectors will be available, and technical and management capacities can be further improved.

8.4.2 Mexican Official Standards, NOM-001-SEMARNAT-1996

This standard aims to ensure that wastewater discharged into watercourses is compatible with the requirements for human and ecosystems use. It stipulates maximum contaminant levels for wastewater discharged into national water bodies and on lands. It represents the basic instrument of regulatory policies for the water quality management of the nation's basins (NOM-001-ECOL-1996). The standard was formulated based on the following considerations:

- Technical feasibility: contaminant removal efficiencies of the wastewater treatment processes and their technological availability;
- Economical rationality: impacts of treatment costs on the financial situation of water utilities, owners and industrial concerns and household consumers;
- Cost reduction in transactions (negotiation, information, monitoring, control and verification) to widen the possibilities of environmental management;
- Gradual compliance: staged improvement with clear, predetermined goals for both municipal and industrial discharges, according to their sizes, population or equivalent pollutant loads;
- Multi-environmental approach: ensure impacts are not transferred from one environment to another, as in the case of by-product sludge;
- Quality of water for specific type of water use: minimum water quality level required by the standard should be suitable for restricted agricultural irrigation, standards become stricter with water uses and types of watercourses;
- Combination of preventive measures and control: reduce pollutants in the productive processes as well as reducing residuals in the source, with control measures for end of the pipe treatment;
- Standards as an instrument to foster development and technological innovation: development of clean and sustainable productive activities also a goal; and
- Provision of a clear framework over the short and medium terms: provide clarity to municipalities and owners of industrial concerns in planning wastewater treatment facilities, with minimum discretion.

8.4.3 Programme for Water Supply, Sewerage and Sanitation in Urban Areas (PROMAGUA)

Sustainable development through proper use of infrastructure is one of the main objectives of PROMAGUA. The actions of this programme are intended to conserve, improve and increase water works, with special focus on such strategic works as large dams and aqueducts. Priority is given to budgetary resources that can specifically address improvements in the quality of water services, including completion of projects in progress and to start new ones in accordance with water demands (CNA/SEMARNAT 2001). Existing projects are improved to enable full-capacity usage. Objectives of PROMAGUA are:

- Help both states and municipalities quantitatively and qualitatively to improve water services in urban areas;
- Support both states and municipalities in improving water facilities, because these levels of the government are responsible for the provision of water services and the operation of infrastructure;

- Gradual reduction of all subsides currently provided by the federal government to water facilities through the CNA; and
- Subsides remaining are oriented to improvements in business and operational efficiency, construction of new infrastructure, especially wastewater treatment plants, and reductions in pollutant loads discharged into water bodies.

Table 8.9 presents total investments in water supply, sewerage and wastewater treatment, for the 1991-2003 term.

Table 8.9. Investments in water supply, sewerage and wastewater treatment (million pesos at constant 2003 value)

Year	Federal	State	Municipal	Other[a]	Total
1991	4,647	3,394	*	3,892	11,933
1992	5,287	2,604	*	2,342	10,233
1993	6,042	3,489	393	2,226	12,150
1994	5,123	1,536	457	1,266	8,382
1995	1,290	1,591	1,023	1,408	5,312
1996	2,184	641	317	93	3,235
1997	2,057	820	809	175	3,861
1998	2,307	612	328	278	3,525
1999	1,949	904	246	196	3,295
2000	2,354	1,464	118	380	4,316
2001	1,116	786	332	647	2,881
2002	1,685	1,006	695	192	3,578

Source: CNA, 2004.
Notes: * These amounts are included under the column named Other.
[a] Investments by state water commissions, land and house developers, loans, USEPA and donations from private partners.

If investments that other federal agencies, such as the Ministry for Social Development (SEDESOL) and the National Public Works and Services Bank (BANOBRAS) are included, the total amount invested in 2002 was 947 million dollars.

8.4.4 Water Rights Tax Rebate Programme (PRODDER)

This programme reflects another strategy to provide total coverage of water services in the country.

After two years of implementation, the total income was 206.7 million dollars. According to the rules under the programme, this increases to 409.1 million dollars, which can be used to reverse the trend of decreasing investments in water projects in recent years. At the end of 2003, the CNA returned 185.5 million dollars to municipal and state water facilities. This fund was expected to contribute to an increase in the efficiency and coverage of water services that will benefit some 4.5 million people.

To participate in this programme, a water service provider must submit an application to the CNA, including a written commitment to construct all wastewater treatment plants, including sludge disposal, in accordance with a detailed programme. The subsequent grants are earmarked for the compliance of wastewater discharges with pollutant levels established in the NOM-001-SEMARNAT-1996 and the Federal Law of Rights.

The programme that has to be submitted considers four stages, and establishes deadlines for their completion, as shown in Table 8.10.

Table 8.10. Stages and terms in the PRODDER

Stage	Activity	Maximum duration (months)
1	Basic engineering	12
2	Executive project	8
3	Construction	24
4	Start	4
	Total	48

Currently, 231 towns and cities of more than 20,000 inhabitants have submitted programmes, and agreements with the CNA have been signed with 224 of them.

Wastewater management in Mexico is governed by federal environmental legislations. The plans and programmes are drawn up using an integrated approach for each river basin, through a consensus that involves the three levels of government and the water users.

Water quality management, however, focuses on pollution due to water discharges, both municipal and industrial, that reach water bodies and lands that are national property. The enforcement of federal regulations regarding discharges is the responsibility of the CNA. However, as mentioned earlier, nonpoint pollution sources, such as wastes from agriculture, livestock, municipal, business and industrial activities also increase water pollution and their control must be discussed at the meetings of the basin councils, where all the main stakeholders participate.

8.5 Reuse

The number of municipal wastewater treatment facilities in the country is 1,242, but only 1,077 are now in operation. In addition, removal efficiency is often not in accordance with design capacity. While the total capacity of the treatment plants is 79.7 m^3/s, only about 56.1 m^3/s is treated. This flow is usually recycled. The national inventory of municipal wastewater treatment plants, as of December 2002, is presented in Table 8.11.

Table 8.11. National inventory of municipal wastewater treatment plants (December 2002)

Federal entity	No. of plants	Operating plants	Design capacity (L/s)	Treated Flow (L/s)
Aguascalientes	91	86	2,723.9	2,230.2
Baja California	15	15	4,437.1	3,897.0
Baja California Sur	16	15	1,082.0	758.5
Campeche	12	12	132.0	45.5
Coahuila	15	6	1,289.0	1,157.0
Colima	44	40	572.0	451.0
Chiapas	16	5	165.3	113.0
Chihuahua	58	56	5,129.1	3,773.7
Distrito Federal	28	28	7,032.0	3,652.0
Durango	89	87	3,431.9	2,337.0
Guanajuato	22	18	3,938.0	2,866.0
Guerrero	25	25	2,861.0	1,656.7
Hidalgo	11	11	102.4	67.9
Jalisco	85	73	2,770.0	2,224.4
Estado de México	59	52	6,616.4	4,550.6
Michoacán	20	13	1,136.0	659.0
Morelos	30	18	1,258.2	1,057.5
Nayarit	56	49	1,660.9	1,092.6
Nuevo León	55	55	12,247.0	8,639.9
Oaxaca	45	37	780.4	595.9
Puebla	32	28	3,101.0	2,320.4
Querétaro	52	48	907.0	622.9
Quintana Roo	14	14	1,536.0	1,021.8
San Luis Potosí	6	5	795.0	545.0
Sinaloa	48	47	3,004.4	2,381.9
Sonora	75	61	3,322.4	2,358.2
Tabasco	39	35	997.5	787.0
Tamaulipas	22	15	2,581.0	2,365.2
Tlaxcala	47	31	920.0	602.7
Veracruz	91	71	2,883.8	1,017.8
Yucatán	11	10	144.5	139.5
Zacatecas	13	11	177.8	160.6
Total	1,242	1,077	79,735.0	56,148.4

Source: CNA, 2003b.

The recycling of wastewater for agriculture began around 1890 when an artificial drain was constructed to alleviate floods damages in Mexico City. Farmers in Tula, Hidalgo, use this water to irrigate cereals and alfalfa. This practice began in the DR 03 irrigation district, and was later adopted by DR 100 in Alfajayucan, Hidalgo.

Other agricultural areas inside the Valley of Mexico also reuse effluents, mainly Chalco and Chiconautla in the State of Mexico. Other places where wastewater is recycled for irrigation include Valsequillo, in Puebla state; Tulancingo, Hidalgo, and Ciudad Juárez, in the state of Chihuahua.

Even though reuse of wastewater for irrigation is a common practice around the country, only a few studies have assessed its effects on the local inhabitants.

Preliminary studies have been completed in areas close to the Mexico City Metropolitan Area. (Tula, Chiconautla, the Lake of Texcoco and Xochimilco), and in General Escobedo, where effluent from Monterrey city is reused (Jiménez and Ramos 1996). It is necessary to conduct long-term studies on the impacts of irrigation with reused water on the human and the environmental health.

Towns and cities whose effluents are reused for irrigation are shown in Table 8.12. It should be noted that only 14 of them provide some form of treatment.

Table 8.12. City effluents that are reused in agricultural irrigation

States	Cities
Aguascalientes	Aguascalientes(1)
Baja California Norte	Ensenada (1) and Mexicali
Baja California Sur	La Paz(1)
Chihuahua	Cd. Juárez, Chihuahua(1), Delicias and H. del Parral
Coahuila	Monclova, Nueva Rosita, Piedras Negras, Saltillo and Torreón (1)
Colima	Colima
Distrito Federal	Distrito Federal (1)
Durango	Durango
Guanajuato	Irapuato (1), León and Salamanca (1)
Guerrero	Chilpancingo and Iguala
Hidalgo	Pachuca and Tula
Estado de México	Toluca
Michoacán	Morelia, Uruapan and Zamora
Morelia, Uruapan and Zamora	Cuautla (1)
Nuevo León	Monterrey (1)
Oaxaca	Juchitán de Zamora and Oaxaca (1)
Puebla	Atlixco, Izúcar de Matamoros, Puebla, San Martín Texmelucan and Tehuacan
Querétaro	Querétaro
San Luis Potosí	Matehuala, San Luis Potosí
Sinaloa	Culiacán, Guasave (1), Mazatlán (1)
Sonora	Cd. Obregón, Navojoa, Nogales (1) and San Luis Río Colorado
Tamaulipas	Cd. Victoria and Nuevo Laredo
Tlaxcala	Apizaco (1) and Tlaxcala (1)
Veracruz	Córdoba, Jalapa (1) and Orizaba (1)
Zacatecas	Fresnillo and Zacatecas (1)

Source: Jiménez B. and Ramos J., 1999

Only two examples have been recently identified of the industrial reuse of municipal effluents. In one case, industrial facilities take wastewater from the sewage system and treat it in accordance with the intended use. Thermal power plants in the Valley of Mexico and Tula, an oil refinery also in Tula, and a metallurgical plant in Monclova, near Monterrey, belong to this category. The thermal power plant in Tula treats 850-1300 l/s and reuses it for cooling. The other case involves a specialist firm, Agua Industrial de Monterrey, that has been treating water and selling it to a small number of companies in Mexico City and Monterrey since 1955.

The national inventory of industrial wastewater treatment plants is shown in Table 8.13, while Table 8.14 presents the geographical distribution of industrial plants in Mexico and the volumes of water they use.

Table 8.13. National inventory of industrial wastewater treatment plants (December 2002)

States	No. of plants	In operation	Design capacity (L/s)	Non-municipal discharge (L/s)	Treated flow (L/s)
Aguascalientes	24	22	197.7	227	70.0
Baja California	181	155	1,000.1	1,283	912.4
Baja California Sur	10	9	189.7	597	189.7
Campeche	46	44	66.2	575	17.9
Coahuila	56	56	1,057.5	1,085	910.0
Colima	10	10	467.0	1,682	314.5
Chiapas	13	11	687.1	2,203	687.1
Chihuahua	22	21	663.3	1,080	287.2
Distrito Federal	3	3	31.14	796	31.4
Durango	18	18	451.7	303	273.3
Guanajuato	56	56	535.2	5,570	235.2
Guerrero	8	7	47.2	657	37.4
Hidalgo	47	47	1,666.8	1,285	1,017.4
Jalisco	54	54	375.2	808	375.2
Estado de México	127	126	1,299.0	2,671	1,032.3
Michoacán	35	33	2,178.5	6,829	1,239.0
Morelos	67	56	850.6	3,291	746.9
Nayarit	4	4	163.0	4,336	163.0
Nuevo León	21	20	3 369.7	2,878	2,233.5
Oaxaca	13	13	869.3	4,817	746.1
Puebla	106	96	601.3	8,583	414.1
Querétaro	90	90	1,323.7	563	513.6
Quintana Roo	2	2	10.5	443	5.0
San Luis Potosí	58	57	524.9	5,525	370.4
Sinaloa	25	21	476.9	42,442	379.2
Sonora	19	17	303.1	22,029	83.1
Tabasco	66	66	429.5	5,206	383.4
Tamaulipas	38	38	1,145.7	6,572	1,041.5
Tlaxcala	70	70	218.5	259	275.9
Veracruz	158	156	12,807.7	34,848	11,102.7
Yucatán	72	62	140.3	1 490	97.0
Zacatecas	8	8	155.8	365	46.6
Total	1,527	1,448	34,304.1	171,298	26,232.0

Table 8.14. Geographical distribution of industrial plants in Mexico

States	Volume used (md³/d)	No. of industrial plants
Aguascalientes	14,519	35
Baja California	864,644	27
Baja California sur	213,256	15
Campeche	1,268,327	15
Coahuila	154,223	31
Colima	6,222,805	13
Chiapas	137,181	64
Chihuahua	49,179	24
Durango	60,279	16
Guanajuato	86,027	62
Guerrero	15,353	21
Hidalgo	76,093	16
Jalisco	358,265	62
Estado de México	400.,85	127
Michoacán	1,498,840	49
Morelos	245,895	46
Nayarit	113,926	9
Nuevo Leon	67,208	27
Oaxaca	569,989	21
Puebla	234,127	65
Queretaro	128,809	44
Quintana Roo	63,139	3
San Luis Potosí	335,999	18
Sinaloa	407,009	50
Sonora	3,202,142	104
Tabasco	519,269	13
Tamaulipas	883,773	56
Tlaxcala	29,786	27
Veracruz	14,712,405	307
Yucatán	16,380	16
Zacatecas	45,428	25
Total	32,944,460	1408

Source: Jiménez B. and Ramos J., 1999.

Non-municipal wastewater volumes reused, by administrative region, with a differentiation between direct and indirect reuse, are shown in Table 8.15

8.6 Clean Beaches Programme

Mexico has around 11,122 km of coasts that offer unique opportunities for recreation, maritime commerce and employment generation for the people living in coastal communities.

Table 8.15. Non-municipal wastewater volumes reused by administrative region, estimated 2002 (m^3/s)

	Administrative Region	Waste-water (generated)	Waste-water (treated)	Direct reuse	Indirect reuse	Discharge to water course
I	Península Baja California	1.9	1.1	0.1	1.8	0.0
II	Noroeste	22.3	0.1	0.0	22.3	0.0
III	Pacífico Norte	42.7	0.4	0.0	41.9	0.8
IV	Balsas	14.8	2.1	0.0	13.8	1.0
V	Pacífico Sur	4.4	0.2	0.0	4.1	0.3
VI	Río Bravo	4.2	3.3	1.0	3.2	0.0
VII	Cuencas Centrales del Norte	2.8	0.8	1.0	1.8	0.0
VIII	Lerma-Santiago-Pacífico	11.1	2.7	0.6	10.1	0.4
IX	Golfo Norte	14.3	1.4	0.2	13.4	0.7
X	Golfo Centro	36.7	11.7	3.4	29.3	4.0
XI	Frontera Sur	7.4	1.1	0.0	5.6	1.8
XII	Península de Yucatán	2.4	0.1	0.0	2.4	0.0
XIII	Aguas del Valle de México y Sistema Cutzamala	6.3	1.2	0.0	6.3	0.0
Nacional		171.3	26.2	6.3	156.0	9.0

Source: CNA, 2004.

Due to the importance of the beaches, and in accordance with the sustainable development criteria adopted by the federal government, the ministries of the Navy, Environment, Health and Tourism, as well as the CNA, the National Fund for the Promotion of Tourism and the Federal Environmental Protection Agency have established a task force. The aim is to apply a combined strategy using both technical and administrative approaches, with a commitment to assess the sanitary conditions of all the beaches, and also estimate their potential for conservation and further development.

Stage one of the project was the launch of a national monitoring system for microbiological water quality that includes 13 resorts and 138 beaches, located in 10 states. The system has recently been enlarged to cover 218 beaches in 33 resorts located in 17 states. It is still too soon to comment on the achievements of the above system.

8.7 Conclusions

While the water quality conditions in several of the country's water bodies are a cause for concern, necessary measures are being taken to tackle the problem. Initiatives such as PROMAGUA and PRODDER are harnessing financial resources for investments in the water sector. Bioindicators are being used as a better way for monitoring water quality. Specific goals and commitments for water users are being established through the issuing of a classification for each watercourse, and specific programmes such as Clean Beaches, are intended to reduce water pollution levels by the combined efforts of the three governmental levels and society as a whole. It is expected that all these programmes will have a major impact in improving the overall quality of water in the country.

8.8 References

Arreguín F (1991) Efficient use of water. Hydraulic Engineering in Mexico 6(2): 9-22

CNA (2004) Water Statistics in Mexico. National Water Commission, Mexico

CNA (2003a) Water Statistics in Mexico. National Water Commission, Mexico

CNA (2003b) National Inventory of Municipal Wastewater Treatment Plants. National Water Commission, Mexico

CNA (2003c) National Inventory of Industrial Wastewater Treatment Plants. National Water Commission, Mexico

CNA/SEMARNAT (2001) National Water Programme 2001-2006. National Water Commission of Mexico, Ministry of Environment and Natural Resources, Mexico

Jiménez B, Ramos J (1999) Potential reuse of water in Mexico. Institute of Engineering, National University of Mexico, Mexico

Mexican Official Standard NOM-001-SEMARNAT-1996. This sets maximum pollutant levels in wastewater discharged into national water bodies and lands. DOF, December 11, 1996

SEMARNAT (2004) National Water Law. DOF 29 April 2004, Mexico

UNESCO-WHO (1978) Water Quality Survey. A guide for the collection and interpretation of water quality data. United Nations Educational, Scientific and Cultural Organization and World Health Organisation. Sydenhams Printers, Poole Dorset, UK

Water Quality Management: Missing Concept for Developing Countries

J. Eugenio Barrios

9.1 Introduction

All the current trends indicate that water quality management (WQM) in developing countries is a very important component of regional and national water-resources management policy and planning. If water management is in its early stages in most of the Latin American countries, WQM is far behind. Several reasons can be noted for this shortcoming, most of which are related to economic constraints. However, these also include poor managerial skills and a lack of understanding of management activities necessary in developing countries. This situation has important social implications in terms of human health and environmental deterioration.

Most of the time, when dealing with quality, water professionals follow traditional quantity concepts and are faced with the problem of limited resources, including lack of trained professionals and the need for expensive equipment and laboratory facilities. This approach becomes an obstacle, because, as in other areas, WQM has usually been applied from the perspective, and on the basis of, models from the United States, Canada or Europe, without considering the main objective: managing water quality within the human and financial resources available in the developing world.

Water quality is a complex issue. Technically speaking, it involves several professional disciplines, such as hydrology, environmental chemistry, biology, social sciences, law, and engineering. It is a stochastic process; that is, it varies over time and space, and uncertainties always exist. The scenario becomes more complex when dealing with WQM, because all the difficulties associated with water management are present, in addition to the contaminant-transport phenomena and all related economic, social, environmental, and health implications. WQM includes all the managerial, administrative and physical measures, either direct or indirect, that are required to control water quality for every specific use and user within a system.

Based on existing information available on water quality, it has become clear that Latin American and other developing countries are not able to afford traditional WQM programmes, which are based on an advanced technical understanding of the issues. New approaches must be formulated, and then implemented, in order to start the management process on the basis of limited, though realistic and useful, understanding. The absence of similar capabilities, as now exist in the industrialised countries, does not have to be translated into a "paralysis-of-action" syndrome. Based on whatever resources that are available, the management process should be launched accordingly, following a learning-curve strategy that is both dynamic and iterative.

9.2 Redesign of National Water Quality Monitoring Network in Mexico

This project was launched in 1996, using part of a World Bank (WB) loan for modernisation of Mexico's national water programmes (WB 1996). The main objective was to redesign the water quality monitoring network, through the formulation of a national programme, component of which were water quantity and quality monitoring, improvement and assessment.

When the project was conceived, it was unique in the sense that it provided an opportunity to accomplish a full redesign of the activity, using new ideas and concepts based on experiences from developing countries in order to develop a functioning water quality information system that would support an efficient management process. Quite a few countries have faced this challenge, because water quality monitoring systems are usually improved incrementally rather than being completely revamped.

The main goal of the redesign initiative was to propose a cost-effective, reliable, accessible, and extensively used water quality monitoring programme (Biswas et al. 1997). The programme included both surface and groundwater.

Specific objectives were as follows (Ongley and Barrios 1997):

- Tailoring the monitoring to specific management and user objectives;
- Reduction and simplification of parameters used for monitoring;
- Optimisation of monitoring sites (locations and numbers);
- Use of a combination of alternate approaches (fixed sites, surveys, etc.); and
- Enhanced efficiency of operations (field and laboratory).

In the case of groundwater, which is very important in Mexico in terms of drinking water supply, the same goals and objectives were followed. However, some adjustments were proposed, such as the development of a preliminary aquifer characterisation (Steele et al. 1997).

The first phase of the project consisted in defining specific objectives to develop a user-oriented programme. First, a consultation process was conducted to identify potential users of water quality information and their requirements, both within and outside the water sector. This proved to be a challenging task, because the activity of the WQM network had always been isolated within Mexico's National Water Commission (CNA) and no interaction had taken place with other governmental agencies at any level. Consultations were carried out with different offices in Mexico's Ministry of Health, Ministry of Environment, Ministry of Tourism, and Ministry of Fishing, as well as with several research and educational entities and the private sector.

The results were somewhat unexpected, because it became clear that nobody had specific ideas about their own information requirements. Data needs were explained in very general terms, more as wishes rather than technical necessities. Ministry of Health staff wanted health-related information, whereas environmentalists and researchers wanted a full description of water quality status in every location and for every day, a logistical and fiscal impossibility. The process, however, was useful in identifying the fact that everybody expected to have whatever water quality information was available, and the importance of proper training and education in the subject was noted.

The survey was complemented by discussions within the CNA to define its specific water quality requirements for management purposes. This proved to be quite difficult. Some of the reasons were related to technical issues while others had to do with the absence of any realistic strategy on water quality within the Commission. In the first instance, the main obstacle was the lack of understanding of the general subject of water quality. The regional and local staff had different backgrounds, such as civil, chemical and environmental engineering, biology, chemistry, or technical support, and they took a traditional, "old-fashioned" hydraulic view of their tasks. The other main obstacle was related to the absence of a specific programme on WQM, which was confirmed later in the process.

After the first phase, a conceptual design was agreed to for the programme. However, it was not possible to define specific requirements in terms of useful information for the management process. The conceptual design was developed, based on WQM responsibilities stated in the National Water Law in 1992.

Due to the absence of understanding of proper WQM, a decision was taken to work closely with the regional and local CNA staff so as to explain and discuss with them the conceptual design before proceeding to the implementation of the redesigned tasks. After visiting the 13 CNA regional offices and working with them, it was clear that it was very difficult for the staff to understand the concept of information generation for management of water resources. The primary reason was that there was no specific WQM policy, strategy, or programme that would let them know what to work for.

The absence of a WQM programme can be seen by referring to the National Hydraulic Plan for 1995-2000 (SEMARNAP 1995), which set only general objectives rather than specific guidelines. The plan's objectives and general strategies for WQM were as follows:

- Conserve water availability and quality for the future;
- Improve knowledge of surface water and groundwater availability in terms of quantity and quality; and
- Increase the economic, social and environmental productivity of water.

In the case of specific programmes and actions, the components were:

- Surveillance of different water quality parameters in the main water bodies; and
- Redesign the water quality monitoring network and reactivate the groundwater monitoring network.

Within the action guidelines, requirements were the following:

- Modernize WQM information systems;
- Promote private sector participation to maintain the WQM networks; and
- Provide special support for controlling pollution in surface water, groundwater and estuaries through the monitoring of biological and physio-chemical quality parameters, in order to assess attainable uses, along with proposed corrective and preventive actions.

It is clear that the interpretation of these guidelines could be quite different form one CNA region to another, and from one professional to another. None of these statements of good intentions had any real meaning when regional or local CNA action was needed. No specific programme translated them into specific actions, in order to define their scope, feasibility, specific goals, and performance indicators. Conceptually, the project was good, but implementation relied on each region's perception of WQM requirements and understanding of its information needs.

The lack of specific objectives, goals and activities, grouped in a WQM programme, was the main obstacle to complete the redesign. The definition of specific parameters and all related technical issues was impossible, as were the procedures to create the necessary information products. Given the complexities of water quality, projects that allow for a broad field of possible actions and understanding made the task very diffuse and ineffective.

After several years, Mexico still remains without a functional, usable and reliable WQM network for its water management activities. The importance of the programme is not in doubt, but its benefits have yet to be demonstrated. As a result, and because of fierce competition among programmes for limited financial resources, its annual budget has been drastically reduced year by year, and the management process continues to lack reliable and useful information on water quality.

9.2.1 Necaxa Basin (Mexico): Water Quality Management Plan

The Necaxa Basin is located in the states of Puebla and Veracruz in central Mexico, about 100 km to the east of the capital. The area of the basin is around 900 km^2, with a population of some 600,000, of whom 80 percent live in acute poverty. The basin includes a system built around 100 years ago, primarily for hydropower generation. It has five reservoirs, with a total capacity of 148 million m^3 and four hydroelectric stations, with a total capacity of 209 mw, built in 2001. The system is operated by a state-owned utility called Luz y Fuerza del Centro (LFC), and the electricity generated is supplied to Mexico City.

Currently, the basin is suffering from most of the problems that affect water resources, the environment, and therefore human health: deterioration because of untreated wastewater discharges which causes a high rate of diarrhoea and similar diseases; lack of solid wastes collection and disposal resulting in garbage ending up in the water bodies; and one of the highest rates of deforestation in the country, producing massive erosion. The reservoirs are losing capacity as a result of high sedimentation, and because of eutrophication, fishing, once a primary local economic activity, has been drastically reduced.

The National Autonomous University of Mexico (UNAM) was asked to conduct a diagnosis on which a proposal for specific measures could be based. An inter-institutional agreement to this effect was signed by the Ministry of Environment and Natural Resources (SEMARNAT), CNA, LFC and the state and local governments.

Several issues that are the focus of this chapter arose during the development of the project. Politics began to take over as each agency tried to define the project on the basis of its own understanding and institutional biases. SEMARNAT wanted an environmental approach, based on forest management principles; CNA wanted the project to evolve as a traditional sanitation programme; LFC was worried about reservoir and water channel operations, and the state and local governments were dealing with the project more in terms of regional development, and as an opportunity to secure additional funds.

The diagnostic study by UNAM showed that all the problems were affecting water quality. This provided an excellent opportunity to use it as the principal management issue, with an additional advantage: each entity was aware of water quality problems and had specific concerns to resolve. LFC wanted to reduce solids and garbage and to protect water quality and quantity; local governments wanted to rescue the landscape and restore fishing capacity in the reservoirs as a development strategy; and the primary interest of SEMARNAT was forest conservation. The main concern of the state government was construction of infrastructure, including landfills and wastewater treatment plants, while that of the CNA was on wastewater discharge control and compliance, wastewater treatment, and flood control.

A potential conflict emerged between the environmental management approach and water management measures that were primarily focused upon sanitation. However, there was a common desire to have an effective watershed management programme. There was no discussion that such a project required an integrated water-management approach (Steele 2004), or even an environmental management programme. However, focusing on WQM would allow the promotion of explicit management objectives over the short-, medium- and long-terms. Restoration of reservoir water quality was a common and key goal for all the participants, and the probability of success was high. Water resources and, specifically WQM, each have a strong environmental component that can be directly applied to the coordination of actions and initiation of appropriate management procedures.

Working under this principle, a management plan was developed using a WQM programme with specific activities, objectives and goals, measured by specific indicators. In addition, an institutional arrangement through a working group within the regional basin council was expected to coordinate implementation of the plan.

Finally, WQM was helpful in combining all the divergent views under a common strategy, where land use, solid wastes collection and disposal, and wastewater collection and treatment were part of the same programme and therefore common to the same objectives and goals. This case is a good example of how WQM programmes can be used to introduce more complex management programmes, such as environmental or integrated watershed management and development, and start to highlight current environmental problems. This is an important issue to consider because managerial skills in developing countries lag far behind what is currently needed.

9.3 Drinking Water Quality Surveillance Systems, Panama

During 2002 and 2003, conceptual designs for drinking water quality surveillance were developed, both in rural and urban areas of Panama (TDS Consulting and Barrios JE 2003 a, b). The rural scheme was the first to be developed, and the urban one followed the same principles. The original purpose of both projects was to develop a surveillance system to assess whether providers of potable water were meeting quality criteria. Existing surveillance systems were quite simple but expensive, and therefore impractical. In addition, information available could not be used to identify real problems, but only lack of compliance involving numerous water quality parameters.

Resources available were limited, in terms both of finance and human capacity. The main challenge was how to develop a quality surveillance system under these conditions and be able to demonstrate the sources of problems. Such a programme implies an integrated perspective on drinking water supplies.

The traditional way of implementing this kind of system is on the basis of water quality laboratory analyses, as the main instrument, supported in some instances by what are called sanitary surveys. However, in Panama, such an approach would have been inadequate, because it would have produced in the best case a surveillance system that could not be afforded over the short- or even medium-term. Accordingly, these projects were developed on the basis of the following principles.

9.3.1 Information System

First of all, the surveillance system should be understood as an information system, where data are generated on the basis of identified objectives and specific needs.

9.3.2 Cycle of Information

Information should be generated following an iterative sequence of activities. This cycle starts with the definition of objectives to fulfil the administrative requirements, and ends with use and application of the information to the management process.

9.3.3 Alternative Water Quality Instruments

In order to reduce the reliance on the more traditional, and more expensive, laboratory water quality analyses, alternative data generation instruments were developed. All of them attempted to use systematically different sources of knowledge, such as user perceptions and common sense, reviews of the inspectors and empirical knowledge, and professional assessments. The methodology was based on survey forms or questionnaires with "yes/no" type questions, specifically designed to prove or discard previously identified potential problems. In addition, traditional technical instruments, such as sanitary surveys and inspections, and essential water quality analysis were also included.

9.3.4 Levels of Information

In order to make information an everyday tool, it was necessary to have different kinds of data, appropriate for each level involved in the management process, from the standpoint of users in Panama's various ministries. The information needs of each level were defined in accordance with the legal responsibilities and specific actions required at different levels.

9.3.5 Drinking Water Quality Management: Source to the Tap

The current consensus is that drinking water quality should be managed from the source to the tap, including the water intake (or wells) and transport, treatment, storage and distribution, and consumption. In keeping with such an integrated approach, it became clear that the primary role of a WQM programme at national and regional levels is to coordinate policies among the different sectors.

9.3.6 Results

The results of these projects demonstrate that it is possible to develop and implement a drinking water surveillance system with alternative trouble-free instruments, if appropriate information system is in place.

The programme translated legal responsibilities and policies into specific actions, to generate information products for different resource management levels. Once water quality information is produced and management actions are taken, the programme by itself will define their own increasing requirements, based on identified information gaps and limitations.

After presenting the surveillance system to different regulatory/management entities in Panama, it was clear that a WQM programme was behind it. The participants in workshops for each system understood the meaning of an integrated water management strategy for drinking water supplies and water conservation. With this strategy, each sector assumes specific roles: source water protection, watershed management, water supply systems operation, treatment and distribution, health impact surveillance, and the ultimate knowledge of safe consumption by the users.

9.4 Water Quality in Latin America

There is a void in the water management agenda in Latin America because WQM has not been adequately addressed in most cases. This situation has been clearly identified in recent water forums.

In 1998, FAO organized an international workshop on WQM and pollution control, in Arica, Chile. One of the recommendations from this event is worth noting:

"Because of the lack of operative and management capacities to develop water quality management programmes, it is important to have professionals able to develop them, at local, regional and national level. A key issue is to make an interpretation of the environmental policies in order to translate them into programme actions" (FAO 1998).

In 2003, the Third Latin American Congress on Watershed Management was held in Arequipa, Peru. It was a well-attended event that included workshops, working sessions and a final declaration by the Congress participants. In the water management group, water quality received special consideration. Among the main conclusions that the participants agreed to by consensus were the following (TCLAMCH 2003):

- Water quality management is a priority; however, there is no clear political willingness to pursue it.
- Lack of reliable information is an obstacle to developing this kind of programme.
- It is very important to formulate and implement national water quality management plans.

9.5 Conclusions

Over the last 20 years, water quality has been receiving increased attention within various water management issues. However, in most cases, the specific management concept is still missing among water and environmental management activities.

One of the main reasons for this unfortunate situation is the poor translation of internationally accepted principles and of politically correct sustainable development water concepts into real world actions. At present, there is no question that the goal is the achievement of sustainable use of water. However, the measures that can lead one country, or even one region, towards sustainability are country-specific. They have to be defined on the basis of local scenarios, priorities and capabilities.

It is very important to note that direct water quality implications occur at local levels, and the level of detail and intensity of a WQM programme decreases as the focus moves to the regional, and finally up to the national level.

Water quality management is an important concept, because it requires that water quantity, pollution control, efficient use of the water, environmental considerations, and human health implications, are integrated as one programme package. Thus, it is a most strategic water management issue.

Water quality management is not a substitute for efficient water management. However, it has to be planned for and developed under its own specific needs, policies, programmes, and actions. Water quality management must be an integral component of all efficient water management policies, plans and programmes.

At present, water quality as an environmental issues is acquiring increasingly more and more relevance, because of the role of water in transporting contaminants and growing concerns over new forms of pollution and their unknown impacts, such as the spread of persistent water-borne pathogens, introduction of genetically modified microorganisms and crops, and endocrine-disrupting substances, in addition to the traditional concerns on persistent organic pollutants and other toxic compounds (NWRI 2001). Regrettably, very few developing countries are now prepared to face these new and challenging concerns.

9.6 References

Biswas AK, Barrios EO, García JC (1997) Development of a Framework for Water Quality Monitoring in Mexico. Water International 22: 179-186

FAO (1998) Recomendaciones a la FAO. Grupo de Trabajo sobre Gestión de la Calidad del Agua. Taller Internacional sobre Gestión de la Calidad del Agua y Control de la Contaminación, Africa, Chile, Septiembre, 30

Ley de Aguas Nacionales (1992) Diario Oficial de la Federación. 1 de diciembre, México

NWRI (2001) Threats to Sources of Drinking Water and Aquatic Ecosystem Health in Canada. National Water Research Institute, Scientific Assessment Report Series No. 1, Canada

Ongley ED, Barrios EO (1997) Redesign and Modernization of the Mexican Water Quality Monitoring Network. Water International 22: 187-194

SEMARNAT (1995) Programa Hidráulico 1995-2000. Secretaría de Medio Ambiente y Recursos Naturales, México

Steele TD (2004) Integrated Watershed Approaches – The 3*M* Concept (*M*onitoring, *M*odeling, and *M*anagement): Block-Course Notes and PowerPoint Presentation Materials. Friedrich-Schiller-Universität Jena, Institut für Geographie, Lehrstuhl für Geoinformatik, Geohydrologie und Modellierung (March 22-April 2)

Steele TD, Barrios EO, Sandoval L (1997) Groundwater monitoring overview: a case study in Mexico. Water International 22(3): 195-199

TCLAMCH (2003) Conclusiones del Grupo Manejo del Agua. Tercer Congreso Latinoamericano de Manejo de Cuencas Hidrográficas, Arequipa, Peru (junio)

TDS Consulting Inc, Barrios JE (2003a) Diseño de un Sistema de Vigilancia de la Calidad del Agua Potable en Zonas Rurales (SVCAPZR). Informe Final, preparado para Ministerio de Economía y Finanzas (MEF) y Ministerio de Salud (MINSA), DISAPAS, Panamá (mayo) 123 p, Referencias y Anexos A-K

TDS Consulting Inc, Barrios JE (2003b) Diseño de un Sistema de Vigilancia de la Calidad del Agua Potable Urbano (SVCAPU). Informe Final, preparado para Ministerio de Economía y Finanzas (MEF) y Ente Regulador de los Servicios Públicos (ERSP), Panamá (diciembre) Referencias y Anexos A-I

The World Bank (1996) Staff Appraisal Report Mexico. Water Resources Management Project Report No. 1155435-ME, Washington, DC

Public Policies for Urban Wastewater Treatment in Guanajuato, Mexico

Ricardo Sandoval-Minero and Raúl Almeida-Jara

10.1 Introduction

The state of Guanajuato is located in the central region of the Mexican Republic. It covers an area of 30,000 square kilometres, or less than 2 percent of the territory of the nation. Some 80 percent of the state lies within the Lerma-Chapala basin, in a zone characterised by the presence of fertile soil in an extensive valley that overlies formations of volcanic rock. Given its location between the country's two principal economic centres, Mexico City and Guadalajara, and its historic role as a necessary staging post on the way north as well as a major mining centre, its economic activity has led to intensive water use. Before the expansion of agriculture that began in the 1930s, this region, known as "El Bajío," was regarded as Mexico's grain-belt. From 1950 onwards, with the development of two irrigation districts and the construction of a power plant and oil refinery as well as the arrival of more efficient pumping systems, agriculture grew on the basis of intensive exploitation of surface and groundwater resources. Meanwhile, there was growth in such petrochemicals-related industries as agro-chemicals and, in the city of León, leather curing and footwear. Between 1950 and 1975, the average annual availability of water dropped from close on 2,800 litres per inhabitant to less than 1,500, a figure that fell to just over 800 litres by the end of the last century. During that second half of the 20th century, the number of deep wells is estimated to have risen from 2,000 to more than 16,000. Such rapid developments have led to growing pressure on the quality of the environment in general, and water in particular.

10.2 Water Use and Balance

Agriculture accounts for nearly 65 percent of the water used worldwide. In Mexico, the figure is closer to 80 percent, and in Guanajuato, it is 87 percent. Guanajuato is a state where agriculture has a major impact on the economy, and agricultural sector is the dominant user of water by far.

In terms of availability, Guanajuato is subject to growing restrictions on access to surface and groundwater. Taking both sources together, average usage comes to 1,400 hm^3 a year more than the renewable volume, including 1,200 that are caused by over-exploitation of the aquifers. Accordingly, groundwater levels are falling by more than two metres in an average year. This intensive exploitation implies an enormous challenge in the future to ensure that adequate supplies are available for use by the population and for productive activities. The challenge is being met through projects that promote efficient use of water in rural and urban areas, promotion of recycling, imports from the neighbouring basins of the Santiago and Pánuco Rivers, and by changing water allocations among different sectors. Table 10.2 shows estimates for the Guanajuato water balance.

Table 10.1. Distribution of water use according to source of supply (percent)

	Agriculture	Industry	Urban and rural population
Surface water	99.7	0.0	0.3
Groundwater	83.2	1.8	15.0
Total	87.8	1.3	10.9

Table 10.2. Water balance (million cubic meters)

Origin	Runoff, Replenishment	Demand, Extraction	Difference
Surface	1,364	1,557	-193
Groundwater	2,949	4,195	-1,246
Total	4,142	5,584	-1,442

Sources: For surface water: National Water Commission, 1993, Hydraulic Balance of the Lerma-Chapala Basin, Guanajuato. For groundwater: Guanajuato Water Commission, 2000, Update of the Balance of Groundwater, Guanajuato.

The intensity of water use in Guanajuato, continuing agricultural water use at greatly differing levels of efficiency, together with the increasing urban and industrial developments, have increased pressures on the quality and vulnerability of both surface and groundwater sources of water supply.

Among these pressures are the following:

- Widespread contamination that arises from agriculture and inappropriate disposal of urban and industrial solid wastes;
- Presence of potentially hazardous activities in areas where the aquifers are vulnerable;
- Potential for the contamination of aquifers caused by land subsidence as a result of over-exploitation of the aquifers or by seismic phenomena;
- Increasing presence of contaminants that are of natural origin, such as arsenic, iron, manganese and lead;
- Higher erosion due to deforestation and inappropriate agricultural practice, leading to the sedimentation of rivers, canal beds and dams; and
- Insufficient control over the discharge of industrial wastes into water bodies at a regional level that affects the Lerma River and its tributaries.

Measures taken to combat these problems include the installation of water quality monitoring networks, mapping of the vulnerability of aquifers, specific attention for areas at risk of contamination, relocation of water supply sources, integrated management of micro-basins, selective afforestation, and an increase in vigilance and enforcement of sanctions for waste discharges. All levels of the government are involved in these measures, but much more remains to be done.

10.2.1 Extent of Service Provision

Table 10.3 shows the extent of the provision of potable and sewer connections in cities and towns of Guanajuato, as well as water supply by means of public hydrants and provision of adequate excreta disposal in rural areas. It should be noted that two-thirds of the state's inhabitants live in urban centres[1]; and of those, almost half live in towns and cities of more than 50,000 inhabitants.

Table 10.3. Extent of provision of potable water and sewage in Guanajuato, 2000

Population range	Localities	With water	With sewage	Total population*
> 50000	12	97.37 percent	95.82 percent	2,213,569
2500-50000	99	96.73 percent	85.89 percent	891,479
<2500	8,821	81.55 percent	39.30 percent	1,520,882
Total	8,932	92.05 percent	75.32 percent	4,625,930

*Total number of inhabitants of private dwellings from 2000 census. Refugees not included.
Source: INEGI, 2002.

It should be noted that almost two-thirds of the population of Guanajuato lives in urban localities of more than 2,500 inhabitants, where access to potable water comes close to 97 percent and to sewage more than 85 percent. At that level, the main problems have to do with the technical and administrative efficiency of the systems, the backlog in repairs and the lack of collectors and treatment plants for wastewater. In the countryside, where one in three of the state's population still lives, the deficiency is greater in potable water and even more so in sewage, access to the latter being less than 40 percent. As a result, the 2000-2006 Administration has changed the structure of the budget from one that was classified in accordance with the source of income to one that is formulated according to the category of expenditure. This change has sharpened the focus and encouraged more intensive consideration of the gravest problems.

Guanajuato has at least five different areas of action for wastewater management, and different strategy is used for each of the following case:

- Towns and cities of more than 50,000 inhabitants;
- Towns of 10,000 to 50,000 inhabitants, as well as areas within the cities that require sanitation solutions on a similar scale;
- Villages of less than 10,000 inhabitants;

---

[1] In Mexico, localities are considered urban if they have more than 2,500 inhabitants.

- Rural communities that have sewage and wastewater treatment systems (normally those with more or less regular water supply systems and a population of around 1,000 inhabitants or more); and
- Rural communities with dry sanitation at domestic level, especially where access to water is very restricted.

This chapter analyses only the first three areas of action in terms of structural factors and current circumstances, as well as strategies for promoting sustained operation.

10.2.2 Extent of Provision of Wastewater Treatment

Table 10.4 sums up the extent of provision of wastewater treatment in 2003, classified in accordance with population size, and the estimated flow of wastewater generated by the sewage systems.

Table 10.4. Extent of provision of urban wastewater treatment in Guanajuato, 2003

Population range	Localities	Flow generated (L/s)	Installed capacity (L/s)	Capacity in operation (L/s)	Effective coverage %
> 50,000	12	6,165	4,165	3,925	63.7
20,000- 50,000	12	880	335	335	38
2,500 – 20,000		361	60	60	16.6
Total	24	7,406	4,560	4,320	58.3

It should be noted that, at present, two wastewater treatment plants account for 49 percent of the installed and operative capacity: León's, with 38 percent of the total, and Irapuato's with 11 percent (Figure 10.1). In contrast with other Mexican states, the existence of a series of medium-seized towns, generally located in relation to the hydrographic grid of the Lerma River and its tributaries, implies an additional challenge in terms of the social, political, financial and operational effort that is needed for their wastewater treatment.

10.2.3 Public Health and Water Pollution

Although establishment of a precise relationship between water pollution and its impact on public health is a question that merits considerable further research, a clear relationship can be noted between the commonest ailments in Guanajuato and the presence of certain disease vectors usually associated with the presence of pathogenic organisms in the water.

The following indicators are noteworthy (State Ministry of Health 2002):

- Infectious intestinal infections was in the 14th place in the overall mortality rankings in 2000; but for infant mortality (less than one year old) and among pre-school children (from one to four years old), it was in third place after "congenital malformations, deformities and chromosomic anomalies," and in sixth place among school-age children (five to 14 years).

- Also in 2000, intestinal infections were the second most common cause of death in Guanajuato overall; intestinal amoebiasis was in fifth place, followed by other helminthiases. Taken together, these three causes accounted for more than 310,000 cases in a year.
- Children are most at risk from this type of illness; after acute respiratory infections, the other most frequent causes of illness among infants, pre-school and school-age children always include intestinal infections, malnutrition and intestinal amoebiasis; the other helminthiases category always appears among the top ten.
- Among hospital admissions, infectious intestinal illnesses account for 2.25 percent of all cases if births and problems associated with them are excluded. Infectious ailments are the cause of 6 percent of hospital admissions, behind only traumas, poisoning, pneumonia and influenza. In 2000, there were 1,510 hospital admissions related to infectious ailments.

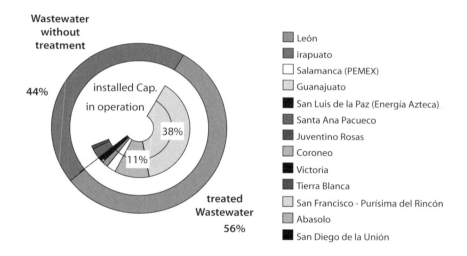

Fig. 10.1. Installed and operating capacity for wastewater treatment, Guanajuato, 2004

While not all of the illnesses and deaths due to intestinal ailments can be attributed solely to the presence of untreated wastewater, it is also true that malnutrition and respiratory problems have some relationship with the use of wastewater in agriculture and absence of adequate mechanisms for the disposal of excreta. At the same time, the presence of certain genetic malformations may be related with environmental pollution, within which the contamination of water and food by wastewater, with a high content of wastes from industry or chemicals used in agriculture, play an important role in Guanajuato. Epidemiological studies have yet to be carried out in Guanajuato that integrate the location and the intensity of the risk factors associated with water pollution and poor disposal practices for solid wastes. Studies are being carried out on the vulnerability of aquifers and establishment of a statewide network of water quality monitoring will provide important information on the basis of which proper decisions can be taken in the future.

10.2.4 Other Harmful Effects of Contamination of Water Bodies

The contamination of surface and groundwater bodies in Guanajuato has the following harmful effects:

- Increase in the cost, or even elimination, of supply sources;
- Proliferation of aquatic weeds that damage aquatic ecosystems and interfere with the operation of structures for the measurement and control of flows, thus increasing the risk of flooding;
- Unregulated irrigation practices with the consequent transmission of disease vectors through food chairs;
- Air pollution caused by the dispersion of particles from wastewater in the dry season; and
- Elimination or reduction of productive activities that require better quality water, in agriculture, industry and recreational activities.

Because of the above mentioned impacts of water pollution on public health, economy, ecosystems, public safety, useful lifespan of infrastructure and the environment in general, including such values as landscape and recreational activities, the state's 2000-2006 Government Plan and the corresponding Water Programme, include a specific target for wastewater treatment and a strategy to achieve it. These measures and strategies will be dealt in the next section.

10.3 Sanitation Strategy of 2000-2006 Programme

10.3.1 Federal and State Regulatory Framework

The federal government, through the National Water Commission (CNA), administers the nation's water resources. The Commission's mandate includes the preservation or restoration of water quality, using the following mechanisms:

- Regulations, in particular those that derive from the General Law of Ecological Equilibrium and Environmental Protection, National Water Law and its regulatory stipulations, as well as by the publication of the Mexican Official Standards (NOMs, according to their initials in Spanish) and, in specific cases, Particular Discharge Conditions. These regulations are based on the definition of limits and conditions for the removal of contaminants and the fulfilment of certain levels within the physical, chemical and biological parameters for the discharge of water into the nation's water bodies;

- Penalties, especially those of an economic nature, as reflected in the Federal Law of Duties and requiring payment of a duty when discharged water fails to meet stipulated standards;
- Coordination, relating to the signing of accords and conventions between the federal government and its state and municipal counterparts in which agreements are made to introduce measures to meet specific targets on the basis of the implementation of financial and technical support programmes; and
- Cooperation, relating to the signing of agreements with users or citizen groups in which special programmes are promoted for meeting the standards.

The first and second cases are of application to water users throughout the country who, in general, are obliged to comply with the legal and regulatory frameworks in terms of fiscal and quality requirements.

The regulatory mechanisms that govern the Guanajuato programme for the treatment of wastewater are the following:

- NOM-001-ECOL-1996. This establishes the maximum permissible limits for the presence of contaminants in discharges of wastewater in the nation's lakes and rivers. According to this regulation, all wastewaters should have been treated in localities with more than 50,000 inhabitants by the year 2000; in those of more than 20,000 by 2005, and in urban localities of 2,500 to 20,000 inhabitants by 2010.
- NOM-002-ECOL-1996, which establishes the maximum permissible limits for contaminants in wastewater discharges into urban and municipal sewage systems.
- NOM-003-ECOL-1997, which establishes the maximum permissible limits for contaminants in discharges of treated wastewater that is recycled for public use.

The *coordination* and *cooperation* mechanisms for Guanajuato relate principally to the constitution of the Lerma-Chapala Basin Council, which was established in 1989 as the first consultative council of its type in Mexico. One of the four original objectives of its programme of accords, commitments and support included wastewater management for the whole basin[2]. Over the years, the basin council's commitments have changed, mainly due to the financial crisis in the years that followed the collapse of the Mexican peso in December 1994. This significantly reduced many of the initiatives that were then underway.

---

[2] The others had to do with matching of availability against demand, efficient water use, and soil and water conservation. This accord is currently under review.

In the context of the Lerma-Chapala basin, commitments were established in three stages:

- In the years that followed the signing of the agreements, the aim was to treat the water supplies for the city of Irapuato and the towns of Abasolo and Santa Ana Pacueco. The first two are the capitals of their respective municipalities; the latter, known as a centre of intensive pork production, belongs to the municipality of Pénjamo. Also included were special initiatives for wastewater treatment for the refinery in Salamanca of the state oil company, Petróleos Mexicanos (PEMEX), as well as encouraging the refinery and other industries of treat their wastewaters.
- In phase two, before 2000, water should have been treated in the cities and towns of León, Silao, Guanajuato, Dolores Hidalgo, San Miguel de Allende, Celaya, Villagrán, Pénjamo, Acámbaro, Salvatierra, Yuriria, Moroleón and Uriangato.
- In the third stage, before 2005, water should have been treated in the towns of Apaseo el Alto, Apaseo el Grande, Comonfort, Jaral del Progreso, Salamanca, Romita, Cuerámaro and Manuel Doblado.

These efforts were scheduled in such a way as to cover progressively all the major discharges which, because of the their flow dimensions and their proximity to the River and its tributaries, contributed to the contamination by wastewater in the Lerma-Chapala Basin. In fact, the programme was not carried out according to this plan. This was because the measures required municipal support, not only in terms of investment but also in political will, as well as acceptance by the citizens of acceptance of higher tariffs. Another negative factor was due to the result of the nation's financial crisis. Two of the original plants, Santa Ana Pacueco and Abasolo, remained non-operational for several years.

Proper implementation of the Official Mexican Ecological Standard 001 that was issued in 1996, would have required an action plan that should have contributed to the following results:

- Before the year 2000, treatment plants should have been built and operational in Irapuato, León, Silao, Guanajuato, San Miguel de Allende, San Francisco del Rincón, Salamanca, Celaya, Acámbaro, Dolores Hidalgo, Cortazar and Valle de Santiago.
- Before 2005, treatment plants should have been built and operational in San Felipe, Salvatierra, San Luis de la Paz, Comonfort, Juventino Rosas, Apaseo el Alto, Villagrán, Abasolo, Yuriria, Pénjamo, Moroleón and Uriangato.
- Before 2010, treatment plants should be built and made operational in Ocampo, San Diego de la Unión, Dr. Mora, San José Iturbide, Apaseo el Grande, Coroneo, Jerécuaro, Tarandacuao, Tarimoro, Jaral del Progreso, Santiago Maravatío, Pueblo Nuevo, Huanímaro, Cuerámaro, Romita, Cd. Manuel Doblado and Purísima del Rincón.

The action plan required by the Official Mexican Standard is determined by the population of the different localities.

Another constraint was the requirement of two action plans, one of the basin council and the other that had to be derived from the standard. Unfortunately, these two plans are not compatible. In addition, both the plans failed to respect the logic that would make it possible for a municipal authority, at any specific point in time, to take measures for the treatment and disposal of its wastewater. The basin council's accords were never fully implemented, nor was the plan required by the standards. As a result, three successive presidential decrees were issued to postpone the implementation in an effort to build up support. The last decree, issued in 2000, offered an amnesty on overdue debts for wastewater discharge rights into lakes and rivers subject to the presentation of a treatment plan that included the construction of sewage collectors and distributors as well as the construction of wastewater treatment plants. Each municipality had to submit its plan, including its components, for approval by the head office of the CNA before signing an agreement with the Commission.

Municipal authorities have finally responded to the most recent decree, primarily because no firm short-term commitment was required and the decree was relatively flexible. This is the general background for the water quality management programme for the 2006 Administration of the state of Guanajuato.

10.3.2 Water Action Programme, 2000-2006

The 2000-2006 water action programme is based on three general objectives:

- Consolidation of participative water management in which joint responsibility is accepted, taking account of the natural environment and ensuring matching between supply availability and demand, efficient allocation of water for public and productive uses, and giving due attention to extreme hydrological events;
- Increase in the efficiency, availability and quality of urban and rural public water services through municipal authorities, and also achievement of similar aims for productive uses; and
- Provision of institutional support for integrated water management through the updating and consolidation of the legal framework, human resources, social values, financial resources and technological developments.

The measures that make up the programme coincide with, or in some way support, the three objectives. For budgetary reasons, however, the treatment of wastewater in urban areas came under the second of the three general objectives: "an increase in the availability of wastewater sanitation systems in urban areas." At the beginning of the administration, it was estimated that an investment of more than US$10 million would be needed to build at least 18 treatment plants which would benefit more than 850,000 inhabitants.

Associated objectives to be met by the construction of the plants were to be:

- Fulfilment of the regulatory requirements;
- Support for the recovery of water quality in the basin;

- An improvement in the health of the population;
- Disposal of the treated wastewater to promote water exchanges with agricultural and industrial users, support for the closure of wells (thus contributing to a reduction in over-exploitation), and the generation of development alternatives based on the water; and
- Support for the recycling of water for uses that do not require potable water within the same system.

The investment that was needed in 2001, for a period of four years, was much more than what had by then been invested by the federal government and the municipalities. In addition, although wastewater treatment plants have become one of the societal requirements, there are several financial, social and technical constraints which still have to be overcome.

10.3.3 Investment Plan for Lerma-Chapala Basin, 2002

As a result of growing concern among the states that form part of the Lerma-Chapala basin, as manifested by various initiatives in the state and federal congresses, a special investment plan for the basin was formulated. One of its components was the construction of sewage collectors and distributors as well as wastewater treatment plants. So far, this is financially viable in terms of the target set by the Guanajuato Water Programme.

CNA delegated to the state commission, in consultation with the national commission's state representatives, full responsibility for identifying measures that need to be taken, investment programme, management of relations with the municipalities and even the feasibility studies and project designs. This delegation of responsibility has contributed to progress in terms of implementation of the programme.

10.3.4 Social and Political Context: Public Acceptance and Demand for Sanitation Services

Ten years of government promotion of water quality management measures, particularly within the Lerma-Chapala basin, together with the growing interest in environmental issues, have created a favourable climate for joint action by municipalities on wastewater treatment. Almost all municipal agendas and programmes currently include construction of wastewater treatment plants. In many cases, citizens are demanding a treatment plant so as to meet regulatory requirements, reduce dangers to health and the local economy and, in some cases, create opportunities for economic development by recycling treated wastewater.

At the same time, in the case of Guanajuato, ongoing work from 1995 to the present with the authorities responsible for municipal potable water supplies and sanitation services, through programmes that improve their technical and administrative efficiency, has reinforced their capacity to meet the financial and operational responsibilities that are implied in the building and operation of a plant.

Interest in "environmental" issues has not been enough to overcome the general resistance to higher tariffs necessary to assure better wastewater management services. In addition, the backlog that has been built up during decades of neglect of wastewater management services means that municipalities do not have the financial resources to build and operate the necessary infrastructure which will comply with the regulations. As a result, the state government has adopted a series of parallel strategies to improve the viability of the sustained operation of wastewater treatment plants and the systems associated with them.

10.4 Elements to Promote Sustainable Sanitation Systems

The Guanajuato action programme for wastewater treatment seeks to create enabling conditions that support the operational sustainability of the systems. These conditions are the following:

- Complementary infrastructure that ensures the capture of wastewater and its transport to the treatment installations;
- Technical studies that characterize the effluents and provide for increasing wastewater generation in the future;
- Estimations, as a part of levels of the tariff studies carried out each year by the state commission, tariffs that need to be imposed due to wastewater management schemes, and proper timing for their application in each municipality.
- Selection of the technology best suited to the physical and economic conditions of each locality, maximizing efficiency so as to minimize operating costs, while taking account of the investment subsidies currently available;
- Explore opportunities for sale of treated wastewater or swaps with the main users so as to reduce the impact of operational costs on tariffs paid by the public and productive users;
- Complementary regulation, especially for discharges to the sewerage systems of the cities where there are industrial activities which produce discharges that are different from normal municipal discharges;
- Identification of the current users of wastewater, irrespective of whether they have the right to do so, and formulation of mechanisms of negotiations with them;
- Reinforcement of the mechanisms for the training and eventual certification of the staff responsible for managing wastewater treatment plants, of which there are to be at least 25 by the end of the administration;
- Support in each municipality for publicity campaigns that raise public knowledge and awareness of the impacts that wastewater may have on health and the local economy, as well as encouraging a sense of responsibility to ensure prompt payment of sewerage charges by consultations (campaigns are to be launched by every operating authority in consultation with users).

- Adequate provision for the disposal or use of sludge generated as a byproduct of the treatment process; and
- Encourage private sector participation in the financing and operation of the systems on the basis of the schemes established by the federal government, like PROMAGUA, and the trust fund for the development of infrastructure, FINFRA.

Some of the measures that are being taken to ensure the fulfilment of the above conditions will be discussed next.

10.4.1 Complementary Infrastructure

In all cases, as recommended by CNA, the requirement is for a minimum collection of 60 percent of wastewater present in collectors or distributors that had been constructed, or in the process of construction. The aim is to avoid the problem that had recurred in previous years when treatment plants were built but remained unused because the wastewater never arrived. Investments to be made in 2003 represented 40 percent of the total budgeted under this heading. This is a reflection of the backlog and the level of investment needed, which would be difficult to pass on as tariffs to the consumers. Once sufficient connections to the sewage system were made, it was possible to extend the programme for the construction of collectors.

10.4.2 Characterisation of Effluents

The feasibility studies include appraisals and sampling of physical, chemical and biological characteristics of wastewater. The principal constraints are budgetary. With limited exceptions, the mainly domestic composition of much of the effluents makes the task more straightforward. The lack of implementation of regulations on land use and/or indiscriminate discharges to sewage systems are the principal risks for the future operation of the plants.

10.4.3 Financial Viability

In terms of financial viability, following three strategies were established.

1. Reduction of levels of investment costs to a minimum. The strategy depends on the flows to be treated. For the smallest plants, the strategy adopted was based on the model established by the state of Nuevo León. Designs were carried out by the Commission itself, thus avoiding royalty payments and the corresponding increase in costs. In the case of intermediate size plants, for which turnkey bids were invited through national tenders, special care was taken in formulating up the terms of the tender to make sure that the bids included all the components of the final product, thus avoiding the inclusion of "phantom" ele-

ments, or omissions that in practice meant extensions that had not been included in the budget. In addition, operating costs were evaluated in detail, especially for energy use. For the larger plants, given the dimensions of the investment and support that were required, private sector participation schemes, as promoted by the CNA, were considered. This was the case for the city of Celaya and for the Salamanca extension.

2. Tariff analysis. Since 2002, the State Water Commission has been analysing the tariffs of the state's 46 municipalities, as well as handling negotiations with the municipal councils at the request of the local authorities, and with the State Congress. As part of this process, the Commission has been including the sewage and sanitation charges in the tariff structure of all the municipalities, and assisting the right time for their inclusion in the bills to be paid by the consumers. The process has not been without political and social difficulties, but progress thus far has been encouraging. Between 2001 and 2004, it has been estimated that, with average tariff increases of 8 percent per year, actual collection increased by more than 50 percent as a result of including new components in billing, which were not previously taken into consideration, i.e., sewage and sanitation charges.

3. Water recycling and exchange. Feasibility studies have shown that it is difficult to sell treated wastewater when the quality and reliability of supply cannot be guaranteed, especially for industry. Several factors constrain recycling possibilities:

- Groundwater is often illegally abstracted without authorisation and payment of appropriate charges, and then illegally sold to farmers by using tankers;
- Existence of recreational users, who at times provide free effluents from their commercial activities;
- Industrial users, who fear losing the titles to their concessions which, in semi-arid zones such as Guanajuato can be worth more than US$1 a year for each m^3 produced. The National Water Law stipulates that when water is not used in accordance with the title to rights, the authority must cancel the rights or assign them to another party to ensure full use;
- Absence of long-term experience for producing adequately treated wastewater;
- Pressure from the original agricultural users of wastewater who want to use the treated wastewater without any charge; and
- Lack of sufficient human resources and effective legal mechanisms to avoid the illegal extraction from surface water bodies, and ensure the effective treatment of wastewater in all discharges from the region.

10.4.4 Complementary Regulations and Agreements with Current Water Users

In the cases of León and Salamanca, the operators themselves have imposed their own regulations for discharges to the sewerage system as well as differentiated tariffs to take account of the sanitation costs. Proper implementation of these measures is the key to ensuring adequate technical and financial operation of the wastewater management schemes.

The issue of previous users of wastewater, with or without rights, is one of the most complex problems found for proper water quality management. In semi-arid zones, with good soil quality and a robust market in agricultural produce, it is virtually impossible to avoid the use of untreated wastewater for irrigation, despite the controls imposed by the CNA. In addition, many agricultural areas, most of them *ejidos* (communal agricultural land), have "presidential endowments" for sewage. In others, it is assumed that existing rights over contaminated water, which were originally introduced for uncontrolled run-offs, now apply to sewage. In 1998, the State Water Commission carried out a study of 12 municipalities in order to identify potential conflicts, since the reaction of society to the elimination of this "source" of supply, whether or not the users have any right to it, is one of the main reasons for the cancellation of the projects.

In all cases, efforts have been made to analyse the potential risks that arise from changes to the discharge locations, flow and quality of water. Rather than seeking the strict application of the law, the Commission decided to negotiate with the users as a means to avoid future conflicts. In the case of Apaseo el Grande, current users of wastewater were recruited in an effort to convince doubters of the benefits of the project. In all these cases, direct interventions by the municipal authorities were essential.

10.4.5 Public Relations

As a complementary strategy, public relations department contact the future operators of the plants under construction to offer them support with promotional materials on the prompt and fair payment of water charges, as well as on water treatment. Other authorities, such as those of León and Guanajuato cities, designed and implemented public campaigns of their own. Given the initial low tariffs, the impact on water bills of the users is both direct and substantial. In many cases, even with the subsidy that has been obtained from the CNA and the state government, the increase on the existing price per m^3 can be as much as 20 to 40 percent.

10.4.6 Training

One issue that has not been resolved is the training in the skills that are required for the sustained operation of the treatment facilities. This includes the ongoing training and certification of competence of the workforce. The strategy has been to include a trial period in the construction contracts of six months, or a year, during which the company that wins the tender has to manage and train the personnel that will be operating the plant. The aim of avoiding very complex technology goes hand in hand with that of not requiring a very skilled workforce, which simply is not available. At the same time, measures taken to improve the technical and commercial efficiency of the operators give them the skills that they will need to run the plants. The implementation of an ongoing certification process will help to provide the necessary continuity.

10.4.7 Disposal and Evaluation of Solid Biological Waste

The sludge generated by biological treatment systems can represent an additional problem. Normally, it is assumed that it will be duly digested and stabilised for deposit in landfills. Three landfills are being constructed to which sludge from the treatment plants can be sent for disposal. One alternative that could be studied is the evaluation of the sludge for the production of compost.

10.5 Progress and Principal Challenges

10.5.1 Measures Undertaken and Planned

Implementation of the programme is being organized in three stages:

1. Between 2000 and 2002, development of feasibility studies (only three treatment plants were completed during this period, Guanajuato, Villagrán, Juventino Rosas);
2. Between 2003 and 2005, construction of the most important series of treatment plants, six in 2003 and a further eight in 2004, together with the rehabilitation of four others;

- New plants during 2003-2004: San Miguel de Allende, Acámbaro, Moroleón-Uriangato, Apaseo el Alto, Apaseo el Grande, two in Cortazar, Atarjea and Santa Catarina;
- Rehabilited plants during 2003-2004: San Diego de la Unión, Abasolo;
- Plants to be built during 2004-2005: Salamanca (second plant), Comonfort, Guanajuato Sur, Pénjamo, Purísima (second), San Felipe, Tarimoro, Yuriria, Valle de Santiago, three plants in Celaya.
- Plants for rehabilitation during 2004-2005: San Francisco-Purísima del Rincón (first plant); and

3. In 2006, consolidation of the operational handover and the financial and operational schemes for each case.

10.5.2 Risks and Strategies

During the process of plant construction, several risk factors have emerged, the most important of which are discussed herein.

- Political changes. New municipal administrations were voted in during October 2003. They knew little, if anything, of the commitments made by their predecessors, had different priorities and suffered from the usual difficulties in managing resources during the first year of their three-year term of office. As a result, they found themselves in difficulties in continuing with their contributions and commitments. Accordingly, the central strategy has been to keep in constant contact with the municipalities and their new management, as well as with opinion-formers and appropriate citizens' groups.
- In the current context, the lack of political continuity can only be addressed by intensive political work by the state government, given that it is sufficiently close to be able to interact directly and regularly with the municipality but is also distant enough to be able to transmit to the municipalities a sense of regional and national priority in conformity with federal guidelines. The political will of the state government, and efficient coordination with the federal government, have been fundamental in this process.
- Lack of financial resources for investment and uncertainty in tariffs. Because of the dimensions of the measures or their complexity, investment in wastewater treatment normally stretches over a time period that goes beyond that of an annual budget. Multi-annual operations still represent certain difficulties in Mexico. At present, a change in the federal government's investment priorities is raising potential problems for the extension of the programme. Mechanisms for the promotion of private sector investment, restricted by the federal programme to towns and cities of more than 50,000 inhabitants, so far includes only two projects: the second Salamanca plant and Celaya (possibly three plants), to which the state government is also being asked to make a non-recoverable contribution. Matters are further complicated by reductions in contributions to the budgets of the municipal governments. Alternatives that are being studied include new financial instruments or obtaining medium-term credits, both of them linked to guarantees on the basis of the participation of the federal and state authorities in the projects. No other guarantee is possible, given that the annual revision of tariffs precludes a stable and foreseeable source of income. Only in two cases, market opportunities exist for the sale of recycled water. In the remainder, the financial situation remains murky.
- Private sector interest. The reform that is needed to give the systems the technical and financial stability has not yet been implemented. This means that broader private sector participation continues to be a generally costly alternative that transfers the most relevant risks to one of the levels of government. In some circles, however, there is considerable interest in achieving the viability of schemes that would allow major private sector companies to take part in projects, particularly in the cities that need big plants. One alternative that could be considered is to open the market in small plants under a strict regulatory framework that would preclude the imposition of additional charges and diversify risk.
- Lack of operational continuity. In Mexico, changes in municipal councils bring with them political pressure to change the management of the operators, who

supposedly are citizens whose terms in office overlap those of the municipalities. Political pressure, however, is difficult to resist. Nor is the re-election of a party in power any guarantee of continuity in a municipality's development policies. Given that the operator is an area of government that manages major cash flows, as well as the political importance of the issue and the possibility that it offers for the granting of "favours," the officials involved are particularly vulnerable to pressure from politicians. Mechanisms for accountability and proper integrated evaluation of the systems' assets, including the intangibles (in particular, the value that derives from the management of its cash flows) could be one way of managing the systems, which are granted to the municipalities by the national constitution.

- Social opposition. The development of projects already under construction can be affected by insufficient care in the implementation of tariffs for sanitation, interference by political or personal groups, vested interests, and other factors. This means that attention must constantly be given to potential flashpoints, by federal and municipal authorities.

- The sanitation situation in other sectors. The existence of discharges close to that of the treated wastewater, whether from other localities, industries or commercial installations, creates the impression in the society of paying for a service that generates no benefits. Given the continued limited capacity for inspection and the imposition of sanctions in accordance with the law, in some municipalities this situation is used as an argument for not discharging the water and reusing it instead.

- Widespread pollution. The benefits expected from treatment of wastewater can be lost if the same area suffers from pollution problems due to leaching of agricultural chemicals, or from landfills. Specialists from an international development bank have questioned the real impact of the Guanajuato programme given the presence of problems that still persist because of such issues. The implementation, however, of simultaneous programmes, such as the federal government's "clean countryside" drive, justify gradual and parallel progress.

- The sanitation situation in other areas of the Lerma-Chapala, Santiago and Pánuco River basins. Success in the integrated wastewater management of a basin depends on comparable progress in its various parts. Otherwise, upstream pollution undoes the clean up efforts of downstream. Equally, downstream contaminants may wipe out the improvements in quality achieved by efforts upstream. Guanajuato is fortunate in that investment in wastewater treatment is planned in cities, such as Querétero and Morelia, from which water comes to the state. In the Pánuco basin, the localities are small, and accordingly their discharges into the River and its tributaries correspondingly minor. A few kilometres downstream the state, the problems are very much greater. Of particular note are the discharges into the River Extoraz, which is to be the source of supply for Querétaro city. The interdependence of the cities may generate a greater commitment so that each one meets its regulatory obligation in terms of wastewater discharges.

- Mechanisms for ongoing coordination to ensure the sustainability of the programme. The lack of continuity affects Mexico's municipal governments every three years, and the state governments every six years. Proposals to provide continuous support include the establishment of a support and supervision area within the State Water Commission itself for the maintenance and operation of the wastewater treatment plants. In plants linked to recycling projects, those who receive the treated wastewater may fight for the continuity of the operation of the plants. If the law remains unchanged but is rigorously enforced, it also makes sense to think that the municipalities will continue their interest in the operation of the plants. However, the best guarantee for continuity and responsibility in the conduct of the municipalities' affairs is participation and direct intervention by their citizens.

- The challenge of cleaning up the environment within the regulatory framework. The schemes chosen for the construction of the plants, for example turnkey projects, in which any technology can be used, leave no time for the negotiating environmental impact permits. The obvious congruence between sanitation measures and improvements to environmental quality ought to be a driving force in the quest for new mechanisms that can make the negotiation of environmental impact permits a more flexible and easier process. In one case in Guanajuato, a paradox emerged when the environmental authority ordered the suspension of construction of a treatment plant because the relevant papers had not been presented. The plant now functions perfectly, though its start-up was delayed, among other things, by the environmental red tape.

10.6 Conclusions and Expectations

Experience in Guanajuato's wastewater management programme has been an evolving process. Consequently, the following conclusions should be considered to be preliminary.

1. A set of factors can affect the success of a wastewater treatment scheme, and they can be classified according to various structures. The factors themselves have been indicated on other occasions. For Guanajuato, these can be classified as follows:

- Water resource: characteristics of the effluent;
- Physical assets: existence of infrastructure and complementary equipment;
- Technology: appropriate selection in terms of operational requirements and financial impact;
- Financial capital: structuring investment and operation in accordance with the cash flows available, in particular level of tariffs and the outlook for the operator;
- Human capital: establishment of a context that favours the attraction and retention of qualified personnel;
- Institutional framework: regulations that encourage adequate charges for the water quality management services, control of discharges into the network, absence of unfair competition for recycled water, clarity and continuity in the

development policies, evidence of equitable fulfilment by all sectors of users and by other areas in the basins; and

• Social aspects: control of potential conflicts because of competition for wastewater and the effective management of social perceptions of the problem.

2. All the above factors, however, will rarely be found in anyone specific context, and thus it would be inappropriate to consider them all before launching a programme, even at a regional level, much less a national one. This means that conditions have to be generated that encourage appropriate investments in water quality management and establish coordination and support mechanisms that are readily available to all the participants, with the aim of overcoming difficulties that can be encountered in the financial, legal, social, technical and political aspects. Flexibility and decentralisation in the implementation of these measures at the level of state government appear, in the Mexican context, to make federal policies effective at the level of municipalities and localities.

The works that are necessary for the collection, treatment and disposal of wastewater are, in the area of hydraulic and sanitation works, among the most complex encountered by municipal public services, taking into account the related social, economic, political and environmental issues. Their complexity merits additional attention.

The definition of a set of criteria and tools that are needed to manage under widely varying circumstances, challenge of implementing sustainable wastewater management systems, and building appropriate capacity in managing such systems, are major challenges that need to be resolved in the future.

10.7 References

CEAG (2001) Situación Hidráulica de Guanajuato. Fortalezas y Retos. Comisión Estatal de Aguas de Guanajuato. Guanajuato, Mexico

CEAG (2003) Programa Estatal Hidráulico 2000-2006. Comisión Estatal de Aguas de Guanajuato. Guanajuato, Mexico

CEAG (2004) Notas informativas sobre situación del programa de saneamiento. Comisión Estatal de Aguas de Guanajuato. Guanajuato, Mexico

CNA (2003) Situación del Subsector Agua Potable, Alcantarillado y Saneamiento. Comisión Nacional del Agua. Mexico

INEGI (2002) Censo General de Población y Vivienda, Instituto Nacional de Estadística, Geografía e Informática, Mexico

State Ministry of Health (2002) Statistics for 2001. Guanajuato State Government, Mexico

The World Bank Participation Sourcebook (1996) Sanitation. World Bank. http://www.worldbank.org/wbi/sourcebook/sbxs01.htm

UNEP (2004) Guidelines on Municipal Wastewater Management. http://www.gpa.unep.org

Wright AM (1997) Toward a Strategic Sanitation Approach: Improving the Sustainability of Urban Sanitation in Developing Countries. UNDP- World Bank Water and Sanitation Programme. http://www.wsp.org/pdfs/global_ssa.pdf

Water Quality Management: North American Development Bank Experience

Raul Rodriguez and Suzanne Gallagher O'Neal

11.1 Introduction

What is so difficult about water quality management? Everyone wants it – governments, water utilities, businesses, the population at large – and yet, it eludes us. What makes it virtually impossible, and why are investors not clamouring to get into the game? There is a permanent and growing demand for good quality water, in fact our lives depend on it. It ought to be a seller's market. It ought to be simple.

However, consider the actual situation. The Camdessus Report notes that $80-100 billion more should be invested annually in water, wastewater and sanitation worldwide over and above the current average of about $80 billion. In the past 10 years, about 5 percent of total private investment and of lending by the multilateral development banks went to the water sectors (Winpenny 2003).

More recently, to put it mildly, private sector investment, which was supposed to be the saviour of the water sector, has stalled. Figures from the World Bank (2004) and Global Water Intelligence (Allison 2002) indicate that in 2002, only 21 municipalities entered into partnerships with private companies for a total estimated at no more than $2.8 billion, hardly coming close to the needs of what ought to be a hot market. The experiment of privatisation in developing countries appears to have been a failure.

Water problems are particularly acute in the poorest developing countries, which at the same time have the fastest growing populations and the greatest amount of catching up to do just to approach the level of more developed nations. Their governments cannot provide the funding needed, nor can they afford to borrow the amounts required.

How, then, does a country address this shortfall effectively and at an affordable price? With a shortage of public funds and the private sector not filling the gap, the water sector is in serious trouble.

11.2 Quality Management

The existing models for water sector management and finance simply have not worked. So, doing more of the same is not the solution. Needs in the water and wastewater sector outstrip the capacity of any government or international finan-

cial institution. The current structure of the water sector inhibits private investment, and it is not even clear that with adequate incentives private investment
would be sufficient. The failure of many of the privatisation efforts has driven
away private capital and commercial banks. Attempting to throw more money into
the same model simply will not work. There is not enough money to meet the
needs.

Before discussing water quality management, it is necessary to consider quality
management. Effective management of the water sector will attract the desperately
needed investment and allocate the resources more efficiently. It is not necessarily
a matter of adding massive amounts of new capital to the sector, but rather of creating a framework that permits optimal use of what already exists. It will be impossible to achieve water quality and extend services until structural changes are
made in how the sector is managed in most of the developing world.

11.3 NADB Experience

When considering management and financing of the water sector, it is worth looking at the North American Development Bank (NADB) experience, not because
NADB has all the answers, but rather because of its specific mandate to lend for
water and wastewater infrastructure – it has experienced all of the vagaries of the
water sector in a compressed fashion. Unlike a commercial bank, NADB does not
have the luxury of selecting only the most solvent clients in optimal conditions to
borrow, nor can it, like other international lending agencies, dilute the risks and
problems to its portfolio through a broader sectoral and geographical distribution
of its loans.

11.3.1 NADB: The Theory

NADB was created as an international financial institution to lend specifically for
water and wastewater projects in the United States-Mexico border region, a narrow, but extensive geographic area in which a developed and a developing country
meet. On the Mexican side, there has been a heavy influx of population driven by
the assembly plant (maquila) industry into an arid region that is not equipped with
infrastructure to attend to this population growth. On the United States side, it is
one of the poorest regions in the country, where thousands live in neighbourhoods
generally unplanned, unincorporated that develop outside the limits of a city without services of water and sanitation.

At the time NADB was created (under the auspices of the North American Free
Trade Agreement, NAFTA), it was thought that water infrastructure needs in the
100 km on each side of the border ranged anywhere from $10 billion to $20 billion. Neither country had the funds available to foot the bill, so NADB was set up
and capitalised at $3 billion (of which $450 million will be paid-in capital) with
the intention of making market-rate loans and guaranties for water projects. The

idea was that the Bank would be able to fund self-sustaining projects and clean up the border. Everyone expected the money to flow when NADB opened its doors. However, this did not happen. What happened?

11.3.2 NADB: The Reality

The estimates of border infrastructure needs were not exaggerated. But needs do not equal projects, much less financially viable ones. NADB quickly ran into the situation that potential lenders and investors face in the water sector everywhere. The difficulties are geographical, political, financial, technical and managerial, as noted in Figure 11.1.

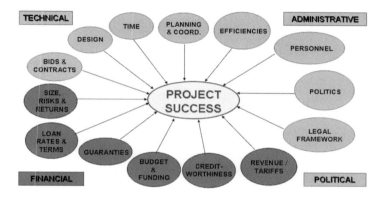

WATER SECTOR PROJECTS
KEY CHALLENGES

Fig. 11.1. Project finance challenges in the water sector, presenting the multiple factors that affect the successful outcome of a water infrastructure project

Any one of the factors noted in the Figure 11.1 can derail an otherwise sound project. Perhaps the project was well designed, but weak procurement capacity resulted in an unqualified firm winning the construction bid. Two years down the line when the treatment plant fails, no one remembers the faulty procurement process that led to poor construction quality: they simply say the project was a waste of public money. It will be used in the next elections to oust the current party, and in the meantime, water service has not improved. In comes the new political party (or even the same party), which promptly removes the head of the water utility and key managers, replacing them with loyal party members. By the time these individuals have learned the water business, it is time for new elections, and still the water is not being treated to adequate standards. The next mayoral candidate accuses the incumbent of planning to raise water rates and wages a furious campaign against the proposed increases.

Having won the election on this platform, the new mayor can either comply with loan covenants calling for increased fees to cover repairs to the sewer lines or refuse to comply because of his campaign promises. The mayor is in a tight spot. When the force main collapses, the utility can only patch what should be a major overhaul of the system. There has been no planning, there are no reserves, and there are no funds in the budget to cover the cost. By the time the planning is complete, the incumbent is no longer in office, and priorities have changed. Back to square one. Variations of this scenario are repeated worldwide in developing countries. Would you invest your hard-earned money under these circumstances?

When NADB opened its doors, it found plenty of compelling needs in the border area; however, there were no projects designed and ready to be funded. Few of the utilities had the capacity to conceptualise and design projects. Even fewer had the credit capacity to take on a loan to finance their water projects. Billing and collection efficiency rates were extremely low so that utilities were not collecting even half of the revenue due to them. Water and sewer systems were full of leaks, causing additional revenue losses. The rates charged to users did not cover routine maintenance let alone capital improvements.

These factors were further complicated by political realities. In Mexico, as in many developing nations, appointment of utility managers is tied to local politics, meaning that management of the utilities turns over every two years on the average, hardly time for anyone to undertake any long-term projects. Compare this with utilities in Texas where the average tenure of utility managers is over 12 years.

In the case of the United States-Mexico border, there is the added complication of transboundary impacts. Water supply issues and wastewater from one side of the border affect the other, at times causing high-level diplomatic incidents.

The legal and regulatory framework in Mexico caused further complications. While managed locally, water is owned at the federal level. Priorities and party politics do not always coincide. Much of the finance for water projects comes from federal transfers, and in times of economic difficulty, funds are scarce.

It is difficult to attract foreign or private capital for projects in this context. The returns are too low. Fees collected from the residents are the only source of utility revenues. In communities with historically very low rates, a politicised rate-setting process, low collection efficiencies and high losses in the systems, projects cannot sustain the cost of financing.

In Mexico, states and municipalities are forbidden by the Constitution to incur debt in foreign currency or from foreign entities. For a few years, this fact alone prohibited the NADB, a Bank created specifically for this end, from lending in Mexico for water projects. It took a lengthy process to establish a mechanism to channel funds from the Bank to the local communities.

In cases where projects were developed, NADB faced additional constraints relating to technical and managerial capacity. For example, introducing wastewater treatment facilities in a community that previously had none requires technical staff with the capacity to operate and maintain facilities, or at least be able to supervise the operator hired to do the work. Internal skills did not exist in the utility beforehand, and with the high rates of turnover in management, it was hard to retain the skills base needed to adequately maintain facilities.

11.3.3 Dealing with Reality

NADB evolved accordingly. One of the first programmes the Bank introduced was the Institutional Development Cooperation Programme. The goal of this technical assistance grant programme was to specifically assist utilities achieve improvements in their operating, managerial and financial operations that would have the effect of increasing their credit capacity. NADB has financed many studies, line surveys and leak detection studies, management and rate studies, as well as billing and collection system studies, with the goal of increasing cash flows and strengthening their financial position. One of the most difficult cases to make to a cash-strapped utility manager is the need to create and maintain reserves. How can a utility leave cash in the Bank when there are so many urgent needs that require immediate attention? It is not an easy sell to convince them that lenders and investors will be more interested in their projects if they have reserves. Yet NADB has seen it happen successfully.

Given the reality of small communities needing large infrastructure investments for the water sector and the impossibility of generating sufficient revenues, NADB worked with the US Environmental Protection Agency (EPA) to create the Border Environment Infrastructure Fund, an infrastructure grant fund that is combined with other sources of finance, including credit, to make projects more affordable to the communities. Without these grant funds, it would have been extremely difficult to launch the lending programme of the Bank. Few, if any, of the utilities would have qualified for a loan, or the projects being undertaken would have had to be scaled back significantly, thereby reducing the impact and improvements.

The most innovative application of the EPA grant funds in projects has been a financing tool referred to as "transition assistance." The Bank has an explicit goal of fostering creditworthiness, funding self-sustaining projects and gradually moving away from subsidies. This means that eventually utilities have to generate sufficient revenues to pay for infrastructure improvements as well as operations and maintenance.

The bottom line has to do with achieving positive cash flows. This must be achieved through a combination of improved efficiencies, cost controls and enhanced revenues. In most cases, they cannot avoid raising user fees, and transition assistance is designed to soften the impact of the increases. Under transition assistance, a utility agrees to raise rates to the level where they ought to be, but the cost to the community is phased-in over several years. The utility raises the rates at a slower rate, and the transition assistance is used to cover the difference between what is actually collected and what should have been collected. Over a period of years, the rates rise in a graduated fashion to the appropriate levels. This has helped utilities meet debt obligations without severe impact on the community.

At the same time, the Bank launched a training programme, the Utility Management Institute, to train utility managers and staff in organisational and financial management, planning and leadership. Managers from both sides of the border are brought together to participate in the training. This has had the multiplier effect of providing training and creating networks of utility directors that consult with each other on management issues. There has been measurable success with the programme. NADB has seen utilities prepare new user-rate schedules and develop capital improvements plans as a result of the training they received, as well as being able to present these concepts to participants in later training sessions.

Technical assistance, training and grants helped to get the initial projects off the ground, but did little to mobilise its capital effectively. With the creation of a Mexican subsidiary, a "non-bank bank"[1] and by gaining access to an exchange-rate hedge mechanism, NADB cleared the way to lend in Mexico. However, the cost of borrowing NADB funds was still unaffordable for the utilities and the communities.

The key modification to the business approach of the Bank has been the creation of a low-interest lending mechanism. The board authorised using a portion of NADB capital to be applied at lower-than-market interest rates.

NADB's portfolio has grown from four projects in 1997, when the only financial tools it had were market-rate loans and guaranties, to the current (April 2004) level of 77 infrastructure projects with a total project value of $2.2 billion.

An additional factor in the business of the Bank is the transboundary impact of two nations at different levels of development sharing groundwater and aquifer sources. Nations with entirely different regulatory frameworks (for example, water is federally owned in Mexico and locally owned in the United States) are using the same bodies of water and sharing the same aquifers, each with different rules concerning drawing water and different regulations concerning disposal of waste. Both countries draw water from the same rivers for irrigation purposes, and this has been complicated by several years of drought. A significant percentage of water is lost through inadequate, leaking irrigation systems. NADB is contributing $80 million of its retained earnings in the form of grants to irrigators on both sides of the border to carry out water conservation projects that will reduce some of the water losses and at the same time, relieve cross-border tensions.

The lesson NADB learned was that it had to adapt to market conditions and to the realities that affect water sector projects in the United States-Mexico border region. This specific approach is not the model for water projects worldwide. Each region and country faces its own circumstances, and the tools developed must be tailored to them. Some themes, however, are common to the water sector itself and can be applied more broadly.

---

[1] NADB created a Mexican non-bank financial institution (Sociedad Financiera de Objeto Limitado, Sofol) that can lend directly to public entities in Mexico.

11.4 Contrast between Markets

Returning to a point raised earlier, consider again the apparent failure of privatisation in the water sector in developing countries. Whose fault is it, anyway? Is it really a case of corporate greed and callous disregard for poor communities? That might make a good campaign sound-bite, but does it correspond to reality? If it is good business and a company is able to make outrageous profits, the last thing the company would do is withdraw from the market. No one has forced the private investors out of the sector, they are leaving because of financial losses and bad experiences.

In this context, another element of the NADB experience has been extremely relevant. Working with water utilities in both the United States and Mexico, NADB can directly compare the experience in management and finance of water projects between the developed and developing world. There are two broad structural differences in the way public utilities are organised and managed and in the types of financing schemes available (Rodriguez 2004).

First, in the United States, there tends to be a firewall between electoral politics and public utility management and municipal management in general. Governors and mayors come and go, but often a city manager and a water utility manager will remain in their positions for years. This provides continuity and experience, the ability to make long-term plans and the timeframe to carry out sustained initiatives that will extend through several local political cycles. This continuity instils a more professional, stable and transparent way of managing municipal governments and utilities. Stability in utility management and good governance foster investor confidence and create an attractive environment for public finance.

Second, there is a thriving municipal bond market in the United States, which is the standard form of finance for water projects. Because of the stability of this market, utilities can obtain financing at very low rates. At the same time, bond finance has a "halo effect" on water utilities, rewarding good behaviour. If a utility does not meet its debt obligations, it will be very difficult to get a good credit rating in the future and corresponding access to finance. Hence utilities are motivated to meet their obligations, set rates at appropriate levels, maintain their systems and plan for future capital improvements. The United States border utilities share the difficulties that are inherent to the water sector itself, but they have the tremendous advantage of a legal and regulatory framework that not only promotes investor confidence, it allows them to operate more like a business.

As is generally the case in more developed countries, United States utilities operate in an environment where contractual obligations are well-grounded, and there is recourse in the judicial system, which provides a level of comfort to bond holders. Municipal bond defaults have been extremely rare, and when they have occurred, the municipality suffered the consequences: public humiliation, the wrath of bond-holders and a significant downgrade in credit rating which increases borrowing costs. Having the appropriate framework underlying the water sector is key to addressing the shortage of capital for water projects.

11.5 Crisis of Governance and Management

The Camdessus Report goes so far as to say that the global water sector is in "disastrous condition." These are harsh words. This situation can also be described as a crisis of governance. The current framework is characterised by:

- Conflicting public policy objectives;
- Politicised and corrupt governance;
- Lack of consensus on the economic value of water;
- Weak legal and contractual configurations and lack of respect for property rights;
- Haphazard regulation and enforcement; and
- Insufficient transparency and accountability, and weak procurement systems.

At the same time, there is a crisis in the management of water systems, in some respects driven by the crisis of governance (Rodriguez 2004), which is manifested by:

- Instability in management and high staff turnover; politicised management;
- Lack of capacity to maintain and develop existing systems;
- Very low operational efficiencies and high losses;
- Deficient pricing practices and tariff structures resulting in dismal revenue streams;
- Poor billing and collection efficiencies; and
- Absence of long-term planning.

The problems in management affect the quality of the system, quality of the water and service and finally, credit standing of the utility, creating a vicious circle from which there is no escape in the current framework. This is evidenced by years of investment that have not rendered the desired results.

The flight of private investors from the water sector points to the conclusion that the conditions for successful investment are not yet present, that the framework is not in place, and that the risks are too high. However, the issue is deeper than whether the sector can attract private capital. The reforms are needed even if the water sector is to be managed primarily by the public sector.

The conflicts around public vs. private sector management have revolved largely around the issue of ownership. Who should own the assets? Should a commodity so vital to the socio-economic well-being of a community be left in the hands of a private corporation? This has led to inevitable squabbles over how much profit is fair, politicisation of fee increases, half-hearted privatisations that are not in the interests of either party, and ultimately bitter recriminations over who is to blame for failed service.

11.6 Systemic Changes and Fostering Investment

A change of approach is needed, moving away from the fixation on ownership to creating a favourable investment climate. When the conditions are right, ownership becomes a side issue. The focus shifts to where the needed funding will come from and how to fill the funding gaps.

This is seen in the developed world where both public and privately owned utilities are able to obtain financing at attractive rates. The focus is on the conditions, not the owner: can the investor expect to be paid, will the water be treated and delivered, are customers satisfied, and what is the track record of the utility? These issues drive investor confidence. This sounds simple, but it is no easy task. Depoliticising the water sector requires a change of human behaviour. It requires the will and determination to do away with political patronage.

Where will the money come from, and how will it be used wisely? These questions are more pertinent than who is the owner of the assets. A new financial structure is required to attract the desperately needed resources for water quality and extension of services, and just as importantly, for operations and maintenance (O&M) of systems. The cost of poor O&M is often more difficult to quantify, yet the impact can be great, in the form of higher costs to operate a system, increased capital needs, and lower customer satisfaction.

One of the key elements is looking to local markets for finance. To the extent that the local markets can be fostered, there will be less reliance on foreign debt, and a strong local market brings benefits far beyond the water sector. It strengthens the economy in general, creating a multiplier effect overall, rewarding financial aptitude, efficiency and good governance with increased access to capital.

Mexico is beginning to launch a local bond market with some success. Municipalities and water utilities are beginning to discover the potential of domestic capital markets as a viable source of funding, which can free them from dependence on scarce grants, budgetary transfers and federal guaranties. In 2001, two state-level bonds were issued for a total of 306 million Mexican pesos. The number and amount of bond issues has grown rapidly. As of February 2004, a total of 24 bond issues had been made by states and municipalities for 11.8 billion pesos. A portion of this capital is being applied to water and wastewater infrastructure. Some of the factors contributing to this rapidly growing market include a stable macroeconomic environment (low interest rates and low inflation), changes in the legal and regulatory framework that allowed subnational entities to enter the bond market, a new practice of obtaining credit ratings that promotes financial transparency, and the availability of national institutional investors such as pension funds and insurance companies with sufficient liquidity to enter the market (Fitch Ratings 2004).

The international financial community can play an important role in the water sector, particularly as market-maker. As far as possible, it should focus on fostering markets and promoting legal, regulatory and institutional reforms to create the right conditions for investment. Development banks can help design the financial instruments, provide support in determining and allocating risks and responsibilities equitably and offer credit enhancements to bolster the local financial markets.

In addition to creating a favourable market, the managerial capacity of utilities must be addressed. Technical assistance and training programmes are needed to build capacity at local level. Reforms at federal level, while needed, are no replacement for strengthening the skills of those actually delivering the service.

There is no single solution that can be applied everywhere. Creating the framework needs to take into account local and regional conditions. The United States-Mexican border is a good example of this. Aquifers cross the border and need to be treated on a regional basis that has nothing to do with state or country boundaries. Water conservation on one side of the border has an effect on the other, while waste discharged into the river affects the other side.

Even with all the most favourable market conditions, financial tools and well-trained staff, the water sector continues to suffer an inherent weakness: the rate of return will continue to be low because the revenue base is the local population. The water sector is typified by a large fixed-asset base and high capital-to-output ratios, high risks and low returns. It is practically impossible to avoid subsidies of some sort. Flat-out grants are not viable, nor are they advisable. They provide no incentives to improve efficiencies or service delivery. A more solid approach to subsidies is making credit more accessible through revolving funds and establishing guaranty, pooling and hedging schemes. These mechanisms would bring down the cost of finance while still promoting good, sustainable practices.

When the public and private sector participate jointly in projects, the focus needs to be more on allocating risks equitably than on ownership. Neither should be saddled with undue risks. Both should be investing so that they have something at stake in the successful outcome of the projects. The recent model of public-private partnerships appears to achieve this more effectively than the previous form of asset sales or concessions. In this structure, the government maintains its oversight and regulatory role, ensuring affordability and access for the poorest members of the community while tapping the capacity of the private sector to mobilize capital more quickly and cheaply and access the latest technologies (Rodriguez 2004).

11.7 Conclusion

In conclusion, it will not be possible to have water quality management until there is quality management of the water sector. Until this issue is addressed in the legal and regulatory framework and in the management capacity at the local level, private capital will continue to search for safer harbours and public finance will continue to fall short of the existing needs.

The political will to change is at the core of structural change for quality management. It is not simply a matter of talent, insight and innovativeness; it is essentially a matter of political determination.

11.8 References

Allison P (2002) GWI Deal Survey Global Water Intelligence. Media Analytics Ltd, Oxford

Fitch Ratings (2004) Mercado de Deuda Bursátil, Opción de Financiamiento para Estados y Municipios. Reporte Sectorial, pp 1-2, México

Rodriguez R (2004) The Debate on Privatization of Water Utilities. International Journal of Water Resources Development 20(1): 107-112

Winpenny J (2003) Financing Water for All. Report of the World Panel on Financing Water Infrastructure. World Water Council, Marseille, France

World Bank (2004) Private Participation in Infrastructure Projects Database. http://rru.worldbank.org/PPI/index.asp

Water Quality Management in Central America: Case Study of Costa Rica

Maureen Ballestero and Virginia Reyes

12.1 Introduction

When talking about the quality of water, it is important to take into consideration not only its physical and chemical characteristics but also to analyse it as an integral part of an ecosystem, and as a finite social and economic resource. This implies integrated management of water for planning, monitoring and protection of soil and water from a sustainable perspective. Based on this philosophy, water quality management requires that the levels of physical and chemical contamination in water for human consumption, as well as in bodies of water in general, should not exceed the permissible limits established by the World Health Organization (WHO).

Deterioration in the quality of water is the result of both natural processes and human activities. The rapid growth of cities and industrial activities such as mining, as well as the intensive use of agrochemicals, have damaged the quality of water in rivers, lakes and aquifers in many Central American countries, thereby further aggravating water scarcity problems. Poor water quality can be harmful to human health and adversely affect crops, water development projects, ecosystems and biodiversity.

Water quality is one of the main requirements for proper management of water resources in the region. The main water pollutants in the seven Central American countries come from untreated domestic wastes, industrial organic materials, pesticides, fertilisers, as well as chemical and pharmaceutical wastes, all of which could have serious effects on health. Other factors include the high concentration of sediments resulting from improper land use practices and deforestation (Davis and Hirji 2003; UNEP 2003).

Central America is especially vulnerable to the effects of extreme and continually recurring hydrometeorological phenomena, such as floods, droughts, and hurricanes, which cause considerable social and economic damages. The most notable such event of the last decade was the Hurricane Mitch, in 1997, which caused widespread damage to infrastructure and crops, with an estimated loss of approximately \$6.1 billion. In addition, changes in land use practices have had repercussions that affect water distribution systems, thus further aggravating the overall impacts of the extreme hydrometeorological events (FAO 2003).

Assessment of the external costs and the environmental damage caused by human activities can be assessed through advances made in environmental and natural resources economics. In general, economic assessments can be made of the magnitude of the damages and its potential impacts on decision-making at economic and political levels and on macroeconomic variables.

Government actions in Central America have generally focused on estimates of the necessary investments for constructing and rehabiliting sewage and sanitation systems, as well as to control discharge of wastes to water bodies. However, reliable estimates of external costs of the social and economic impacts of water-borne diseases are not available at present. Instead of merely focusing on the end results, both short- and long-term benefits would be generated by the formulation and implementation of preventive rather than corrective and/or remedial policies. Further, economic and social benefits can be generated by using good practices or by the implementation of cleaner production programmes.

Given the informational needs of the decision-makers, this chapter analyses the issues surrounding water quality management in Central America through a critical analysis based on a national case study. Costa Rica was selected because it reflects the situation throughout the region, given its fragmented institutional regime for water resources management and dual policies relating to potable water and sanitation.

This chapter analyses specific sectors such as water treatment and sanitation, as well as investments made in the sector and their relation to the Gross Domestic Product (GDP) and the national budget. The state of water quality in Central America is summarised in the next section. It is followed by the case study of Costa Rica.

12.2 Water Quality in Central America

With a territorial area of 521,598 km^2 (2.6 percent of Latin America), Central America contains 23 international river basins and a population of 37.4 million (PNUD 2003). In general, the volcanic range, which extends through the region from north to south, serves as a divider between the Pacific and Atlantic subregions. The distribution of water resources, both over space and time, is irregular in the region. The Atlantic Coast is generally more humid than the Pacific. Depending upon the distribution patterns of water as well as its demand, many territories and populations suffer from regular water stress[1]. In Central America, levels of stress vary from 0.6 percent in Belize to 5.1 percent in Costa Rica (UNEP 2003).

The availability of water for human use and consumption should be evaluated as a function of both quantity and quality. The quality of potable water is expressed in terms of permissible limits in terms of physical, chemical, biological and radiological parameters found in water. The comparison between the values

---

[1] The idea of water stress is applied by using two different parameters. One is water shortage per inhabitant. In this case, there is shortage if water availability is less than 1700 m^3 per person per year; severe shortage means less than 1000 m^3 and total shortage less than 500 m^3 (Falkenmark and Widstrand 1993; Ohlsson 1998, in UNEP 2003).

measured with these permissible limits (norms or standards) determines whether water is potable or not.

From a health viewpoint, access to safe water, and proper disposal wastewater, are essential for the control of a number of infectious diseases. Illnesses that are transmitted through water, such as cholera, typhoid, viral hepatitis and diarrhoea, are closely linked to the consumption of contaminated water. The population most vulnerable to the consumption of unclean water include children, elderly, and malnourished people, since their defences are lower than those of the normal healthy adults. Diarrhoea alone accounts for almost 25 percent of mortality in children under five years (PAHO 1996, in UNEP 2003).

The Central American countries are at a similar stage of development in terms of water coverage for human consumption and sanitation (OPS, OMS 2000a-f). Conditions may vary in different countries in accordance with political, social, economic, environmental and institutional conditions. The total coverage of potable water in the region is 78.6 percent, of which 93 percent corresponds to urban areas. Even though this percentage has increased in recent years, it has sometimes been accompanied by deterioration in the quality of services received. In Belize, Costa Rica, Nicaragua and Panama, coverage in urban areas is expected to reach 100 percent through a policy of disinfection. However, water disinfection coverage in urban areas of Honduras reaches only about 50 percent, and in Guatemala just under a quarter.

In some countries, access to water has improved significantly in recent years. In Belize, almost 100 percent of urban areas are covered and 82 percent of rural areas. In Guatemala, irrespective of access to clean water of nearly 99 percent in urban areas and 70.3 percent in the countryside (1999), the overall quality of raw water, in terms of its physical, chemical and bacteriological qualities, is unacceptable for the majority of surface water sources. Likewise, Honduras is one of the most critical cases: the provision of disinfected water for domestic consumption is around 45 percent (OPS, OMS 2000 a-c).

With respect to sanitation, the situation is even more alarming. Almost all the countries have serious problems with the disposal of wastewater because most of sewer systems discharge directly to water bodies without adequate treatment. The average sanitation coverage in the area is 77 percent (91 percent in urban areas), whereas only about 50 percent of the population have access to sewage systems and only 24 percent receive some type of treatment, one of the lowest in Latin America. OPS and OMS (2001) indices indicate that countries such as Panama have systems that provide water supply and sanitation coverage for nearly the entire urban area, while the coverage in Belize does not even reach 40 percent. These estimates, however, include the urban population who are connected to the public system as well as those linked to septic tanks or latrines.

The main illnesses transmitted by water that affect the population in general, and children in particular, are diarrhoea, hepatitis and cholera. El Salvador has high indices of infant mortality because of serious water contamination, estimated in economic terms at between 1.2 and 1.7 percent of GDP. In Nicaragua, between 1992 and 1999, approximately 32,000 cases of cholera and almost 2 million cases of diarrhoea were recorded. Another widespread illness related to consumption of unclean water is viral hepatitis. In 1998 alone, more than 1000 cases were reported.

The region suffers from a low coverage of water services in rural areas, as well as by intermittent services that often leave the population without water supply for several hours at a time, along with low levels of coverage of the sanitary systems in both rural and urban areas. For example, in Panama City, all the sewers discharge directly into the bay. Furthermore, the absence of appropriate institutional framework for regular monitoring has led to a reduction in the quality of water in both urban and rural areas. Water quality has also been affected by the contamination of water bodies, as in El Salvador, where 90 percent of river waters are estimated to have been in some way contaminated by domestic, industrial, agro-industrial and medical wastes (OPS, OMS 2000d-f).

Water quality problems in the region have been increasing throughout during the last decade. These include wastes generated by industrial, agro-industrial and farming and fishery activities, all of which discharge their waste products into water bodies, in most cases without any treatment. Industrial activities are estimated to contribute to about 44,000 tons of organic contaminants per year, which are discharged into the rivers of the Pacific. It should also be noted that industry is the second largest user of water resources in the region (PNUD 2003).

One of the major factors affecting the quality of water for human consumption and the coverage of sanitation in the region is the lack of investment. It is estimated that an investment of $3.06 billion, approximately 5.8 percent of the GDP, is needed to satisfy the demand, or the basic needs, of the majority of the population in Central America. The investment required to meet the demand for potable water in the region is estimated at approximately $1.48 billion, along with $1.58 billion for sanitation. The most critical cases are those of Honduras and Nicaragua, which require an investment of $95 per inhabitant, followed by Guatemala ($87.2) and Costa Rica ($78.4). Nevertheless, Guatemala is the country which needs the largest total investment of $1.038 billion (6.3 percent of the nation's GDP) (BID 2003).

In general terms, the region has advanced considerably during the past decade with respect to the distribution of potable water. Nevertheless, it is still facing considerable problems in terms of water quality, with the most critical countries being Honduras and Guatemala. Proper systems for monitoring and controlling water quality for human consumption in the region need to be installed on a priority basis. Regarding the sewage system and sanitation, practically all the countries face critical situations which require immediate priority. Poor water quality conditions have resulted in the proliferation of illnesses such as diarrhoea, hepatitis and, in some cases, cholera. Through the analysis of the Costa Rican case study, this chapter will outline how the water quality problems of the Central American region can be addressed.

12.3 Case Study: Costa Rica

Costa Rica is a small country, having a territorial area of 51,100 km^2 (9.8 percent of the total area of Central America) and 3.8 million inhabitants. It has an average population density of 74 persons/km^2, the greatest concentration being in the Central Valley, with more than 1000 persons/km^2. Costa Rica has a 4.8 percent level of illiteracy, with 96.8 percent of the population having access to electricity (INEC 2000). It has one of region's highest national income per capita, at $4,029 (BCCR 2004).

According to the risk and regulation regimen methodology developed by the London School of Economics, the management of water resources in Costa Rica, as in the rest of Central America, is characterised by the absence of clear policies, outdated legislations and the lack of, and/or overlaps, in capacities and functions among the leading public, private or external entities, supervisors and executors[2]. The water resources administration is highly fragmented, with several entities in charge of carrying out any specific function.

The Ministry of the Environment and Energy (MINAE), in accordance with the Environmental Law (Law 7594), has the responsibility for policies on water resources in terms of the protection and maintenance of their supply. MINAE is thus in charge of granting concessions for water use and permits to drill wells. It is also responsible for overseeing complaints about contamination and environmental damages, along with the Agrarian Court. In addition, the Ministry of Health (MINSA), according to the General Health Law (Law 5395), is responsible for establishing the technical norms with respect to the quality of potable water, surface waters and treatment systems.

Both institutions, to a varying extent, are in charge of regulating and supervising supply and demand in terms of the quality and quantity of both surface and groundwater. With regards to water quality, other organisations also have responsibilities. These include the Costa Rican Institute of Water Supply and Sewage Systems (AyA), Costa Rican Social Security (CCSS), Company of Public Services of Heredia (ESPH), municipalities, Committees of Rural Water Supply, (CAARs) and Administrative Associations for Water Supply and Sewage Systems (ASADAS), among others. In addition to these institutions, there are several legal and judicial instruments such as the Water Law (No. 276) of 1942, General Law on Potable Water (No. 1634) of 1953, General Health Law (No. 5395) of 1973 and the Regulating Decree for the Quality of Potable Water (25991-S) of 1997.

With respect to policies related to the quality of water, in 2003, the Ministry of Health (MINSA), Costa Rican Social Security (CCSS), Costa Rican Institute of Water Supply and Sewage Systems (AyA) and National Insurance Institute (INS) came up with a Concerted Sanitary Agenda in which they made ten commitments for the sector for 2002-2006. Item number six of the agenda considers the im-

---

[2] Reyes et al. (2003) Régimen del Recurso Hídrico: El caso de Costa Rica. GWP-Centroamérica-CINPE-UNA. San José, Costa Rica.

provements in environmental health, with special emphasis on basic sanitation and integrated management of water resources.

The country has advanced considerably in terms of the protection of water resources through the implementation of a programme involving payments for environmental services, as well as in the coverage of water for human consumption, which are analysed later. However, the quality of water for human consumption has been neglected, as have the sanitary and control systems of the water supplies generated by different economic activities.

The main sources of contamination of surface waters are:

- Improper sanitation, due to inadequate sewage systems;
- Uncontrolled urban development;
- Accidental spills;
- Industrial discharges without proper treatment; and
- Excessive use of agrochemicals in agricultural activities.

Groundwater contamination is mainly caused by:

- Discharges from septic tanks;
- Agrochemicals from farming activities;
- Faecal contamination from domestic wastes;
- Salt water intrusion in coastal areas;
- Industrial waste discharges; and
- Reuse of wastewater in irrigation.

With increasing social concern on pollution, a National Programme for Improving the Quality of Water for Human Consumption in Costa Rica (PNMCACH) was proposed (Mora and Portuguéz 2000) for the period 2002-2006.

12.3.1 Quality of Water for Human Consumption

In Costa Rica, 97.4 percent of the people have access to water in their homes, with 76 percent having access to potable water. Some 89.3 percent of the population is supplied by water supply systems, 5.5 percent from wells, and the rest obtain their water from rivers, creeks or other natural sources, including rain. Ninety-seven percent of the population have indoor plumbing and 89 percent have sewage or septic tanks (INEC 2000). The main driving forces for improving coverage and water quality are organisations such as the Costa Rican Institute of Water Supply and Sewage Systems (AyA) and the Public Services Company of Heredia (ESPH). While the progress made in recent years is significant, it still is inadequate in terms of good public health. AyA has been a pioneer in establishing water quality control programmes for human consumption, but much more remains to be done.

The National Water Laboratory (LNA 2003) of the AyA is working on the development and implementation of a monitoring programme. Through this programme, the levels of faecal contamination present in the water supply systems are identified, and water provision systems are subjected to sanitary inspections. The main objective of this programme is to reduce the risk to the population due to possible sources of contamination. The programme includes monitoring the water of water supply systems by using samples obtained in accordance with the proportion of population supplied, close to 58 percent. Controls are based on WHO (1984) guidelines. Analyses are carried out at source, in the storage tank, and in the distribution networks. This is done daily, monthly or quarterly, depending on the number of population supplied and the type of water sources (surface or groundwater).

This programme is particularly innovative and useful because in the past only bacteriological monitoring was carried out only at certain times. In contrast, the current monitoring considers the historical backgrounds of water supply systems and potential points of risk. Although the national health indicators demonstrate significant advances in recent years, there has nevertheless been an increase in the occurrence of diarrhoea incidents due, to a large extent, to the deficiencies in the quality of water supplied to certain sectors of the population. Overall, of the total of water supply systems in Costa Rica, only 48.3 percent provide water of potable quality, and only 41.6 percent of the population receive chlorinated water in 2002.

According to data available form the National Water Laboratory in 2002, some 170 water systems administered by the AyA were monitored, as were 1,901 administered by other organisations. Of the total 2,069 water supply systems monitored, 48 percent were found to provide water of potable quality. Only 33 (1.6 percent) had water treatment plants, mostly in medium-size or large urban areas, while 20.6 percent were found to provide disinfected water (Proyecto Estado de la Nación en Desarrollo Humano Sostenible 2002). The most severe contamination problems appeared in about 200 systems which supplied water from surface sources, such as rivers and creeks, without any treatment. These systems are vulnerable to increases in turbulence and contamination caused mainly by the discharge of domestic and industrial wastes and soil erosion. Examples of these include the Virilla, Bananito and Liberia Rivers.

Chlorination of water for human consumption has been one of the most significant advances made in public health on a global scale. Although established as a norm during the 1960s in Costa Rica, only 18 percent (363) of water supply systems are chlorinated (LNA 2003). These are mainly those that are operated by AyA and ESPH. Despite notable increases in the number of chlorinated or disinfected water supply systems between 2001 and 2002 (19.8 percent and 20.1 percent respectively), only 416 of them have continuous disinfection. As a result, only 78.4 percent (3.9 million) of the population are supplied with potable water, while the remaining, 21.6 percent (0.9 million), are in a constant danger of drinking non-potable water.

Water quality deterioration and water scarcity are directly related to population increases and economic growth of the country. It has become increasingly difficult and complex to satisfy the escalating water needs of the society, both in terms of quantity and quality. Water treatment requires diverse processes, but the minimum required is the elimination of viruses, bacteria, and parasites.

Methods of water treatment that are most commonly used in Costa Rica are disinfection and use of slow and fast filtration treatment plants. During the disinfection process, chemical substances such as chlorine are added to the water to eliminate microorganisms. Despite reasonable coverage for human consumption, operators have yet to meet the minimum technological requirements for the provision of good quality water. The situation is further complicated by the failure to get all the treatment plants properly operational. According to the Department of Studies and Projects of AyA, most of the treatment plants that are not administered by AyA are either not functioning or face serious structural problems (Fallas 2003, personal communication). AyA administers approximately 35 treatment plants. The use of substances such as aluminium sulphate and gaseous chlorine can guarantee that the water will be of good physical, chemical and microbiological quality.

The cost of the chlorination depends on the mechanisms used and the population supplied by the system. Thus, for example, a system of sodium hypochloration by electrolysis that supplies a population of approximately 600 consumers costs around of $2,500-3,000, while a system using chlorine gas for the same population would have an initial cost of $5,000-6,000. It should be noted that investment in water treatment systems alone is not sufficient to ensure supply of good quality water. Strict control of water quality is also necessary, as well as a good monitoring programme, experienced and committed technical personnel, and a civil society that values and appreciates the importance of good quality water as an indispensable requirements for sustainable development.

12.3.2 Wastewater and Sanitation

Even though Costa Rica has legislations and regulations for the disposal of wastewaters, in practice it is not always enforced. The direct discharge of wastewaters into rivers and the excessive use of septic tanks are two important factors that contribute to the contamination of surface and groundwater bodies. Some 250,000 m^3 of sewage are discharged daily directly into the Virilla River, tributary of the Grande of Tárcoles River. Along with the Reventazón River, the Virilla River receives 70 percent of all the untreated wastewaters of the country (La Nación 2004). The inefficient handling of wastewaters is due to technical deficiencies in the treatment systems, excessive use of septic tanks, absence of resources for developing integrated measures, and the education and customs of the population, as well as to overlapping responsibilities of the institutions involved. AyA, for example, is responsible for the wastewaters in the sewage system until they reach the treatment plants. Thereafter, control of the direct discharges of water into other natural channels is regulated by the Ministry of Health.

The Regulations on Discharge and Reuse of Wastewaters establish the minimum parameters that should be sampled, as well as minimum frequencies for sampling. Operational analyses and reports should be submitted to the appropriate institutions according to the specific case. For example, if the runoff is to be discharged to a sanitary sewage system, the results have to be reported to AyA; while runoff discharged into a natural channel, must be reported to the Health Ministry. Recycling of wastewaters is acceptable only when there is no deterioration in the quality of water, either surface or underground. Recycled water is used in the rural areas mainly for irrigation, recreation and construction.

One of the causes of inefficiency in the handling of surface runoff is the excessive use of septic tanks in Costa Rica. Some 70 percent of the population use septic tanks for the treatment of domestic wastes. Many of these systems have serious structural and operational problems, and it is quite common to see leakages from septic tanks on the streets. Deficiencies in septic tanks often mean that homeowners release sewage onto the land. Professionals involved in construction, as well as the users themselves, frequently have no idea of the importance of complying with the necessary regulations. The operation and maintenance of these systems are frequently ignored (Céspedes 1993). Soil studies are often not carried out prior to their construction. Tanks are often constructed without any knowledge of the phreatic level, resulting in faecal matter being infiltrated into the aquifers (Reynolds 2002).

Another type of treatment system for domestic discharges involves treatment plants. Table 12.1 includes a list of treatment plants by type, municipality and administrative body. Studies by AyA, in 2000, demonstrated the ineffectiveness of systems located in the Greater Metropolitan Area (GAM). Most of the treatment plants were found to be in a state of abandonment, while few others were functioning but only minimally. Many lack a manual for operation and maintenance, in addition to which they discharge wastewater directly into the rivers. Lack of maintenance has meant that structural damages often go unrepaired. Sludge produced in the treatment processes is not removed, thus causing blockages and overflows. Wastewater is often discharged directly into rivers and streams without the use of any treatment process, as in the case of industrial plants of San Sebastián Rey, La Florita and Cerámica Poás (Araya 2000).

There are some treatment plants outside the GAM, most of which were constructed more than 20 years ago. They are already obsolete. The stabilising lagoons of Liberia in Guanacaste are a good example. When they were constructed, they were efficient, removing more than 98 percent of polluting agents with two stabilising lagoons (Céspedes 1993). Now, however, even though there are four lagoons treating wastewaters, many of the city's houses are not served by the sanitary sewage collection system. Nor does the system has the flexibility to absorb the continually increasing population growth. Consequently, the Liberia River has a high level of faecal contamination, with average values of 1,020 to 88,421 coliforms per 100 millilitres, (Mora and Portuguéz 2002). This is because untreated domestic waters are discharged directly from the stabilising lagoons. This, in turn, means that the water is not suitable for irrigation, recreation and fishing (Mora and Portuguéz 2002). It also increases the risk of water-borne diseases among the population.

Table 12.1. Type of treatment plants by province, municipality and administrative body

Province	Municipality	Administration	Type of plant
San José	19 Pérez Zeledón	AyA	Stabilising lagoons
Alajuela	01 Alajuela	Municipality	TI
Cartago	01 Cartago	Municipality	TI
Heredia	01 Heredia	ESPH	TI
	01 Liberia	AyA	Stabilising Lagoons
Guanacaste	02 Nicoya	AyA	Stabilising Lagoons
	03 Santa Cruz	AyA	Stabilising Lagoons
	06 Cañas	AyA	Stabilising Lagoons
Puntarenas	01 Puntarenas	AyA	LAC

Source: Araya et al., 2003.

12.3.3 Sanitary Sewage Systems

Costa Rica lacks an efficient sanitary sewage system. Efforts made decades ago collapsed because of increases in population and poor operation and maintenance practices. Only 24.8 percent of the population are covered by the sanitary sewage system. Not all of the collected wastewaters reach treatment plants. Only one out of every 15 inhabitants is served by sanitary sewage systems and treatment plants, and the percentage of population that served at present by a sanitary sewage system and a treatment plant that are actually in operation is even smaller, 2.4 percent. Table 12.2 shows that population covered by sanitary sewage systems by provinces. The table shows that the greatest coverage can be found in San José (51 percent), followed by Heredia and Cartago (15 percent). In the remaining regions, coverage is much less, around 10 percent.

Treatment and disposal of wastewaters are currently best in the Greater Metropolitan Area (GAM) and in the cities of Liberia, Santa Cruz, Nicoya, San Isidro de Perez Zeledón, Puntarenas and Limón, whose sewage systems and treatment plants are administered by AyA. These plants do not always cover the total population, since sections of the population are located downstream of the treatment plants, as in Puntarenas, or because the maximum capacity for which the plant and the sewage systems were designed have already been exceeded, as in the stabilising lagoons of Liberia (Araya 2001) and San Isidro del General (AyA 1998).

Table 12.2. Population covered by sanitary sewage system per province

	San José	Alajuela	Cartago	Heredia	Guana-caste	Pun-tarenas	Limón	Total
Total Population	1,345,750	716,286	432,395	354,732	264,238	357,483	339,295	3,810,179
With sanitary sewage system and treatment plant								
Served Population	36,716	48,351	59,126	53,296	26,015	25,830	3,210	252,544

Percentage Coverage	2.7	6.8	13.7	15	9.8	7.2	0.9	6.6
With operational sanitary sewage system and treatment plant								
Served Population	15,113	2,292	6,142	14,256	26,015	25,830	246	89,893
Percentage Coverage	1.1	0.3	1.4	4.0	9.8	7.2	0.1	2.4

Source: Araya et al., 2003.

In the GAM, 68 percent of the inhabitants are connected to one of the four existing collectors[3] that gather the wastewaters and carry them directly downstream into the Tárcoles River. In some cases, construction companies are required to build the sewage system up to a certain point in the network, even though the sewage system is not operational. This is done in anticipation of the fact that some time later these can be connected to a wastewater treatment system, when it is completed.

The existing situation indicates that the wastewater management system is generally not functional, in contrast to the system of water supply for human consumption, which, although in need of improvement, still covers most of the population. As the organisation responsible for potable water and the sewage system, AyA has been quite efficient in providing potable water. However, the situation is now critical with respect to the management of wastewaters. For the sewage system alone the investment needed to cover the deficit is estimated at approximately $157 million (BID 2003).

While monitoring and control will not automatically increase the quality of water for human consumption, they are essential tasks for management purposes. Such a process will ensure timely identification of water quality problems so that public health is not at risk. Although the various institutions operating the water supply systems have made efforts to improve the quality of the water that they provide to the population, they are many areas where additional investments are to be made to further improve water quality. At present, there is no monitoring for heavy metals or agrochemicals, and many other industrial waste products. The efforts of the Ministry of Health to disseminate information and monitor water quality on behalf of AyA need substantial improvement. A water culture is needed that values the resource and involves the sharing of responsibility between society and the state institutions.

---

[3] The four collectors for the GAM are: 1. Rivera collector (the main sub-collectors of this system are Barreal, San Blas, Tibás, Zetilla and Los Colegios); 2. Torres collector (main sub-collectors are Saprissa, Negritos, Cangrejos, Lantisco and Psiquiátrico); 3. Mo Aguilar collector (with the contribution from the sub-collectors of Purrures, Ocloro, del Sur, Las Arias, Chile de Perro), and 4. Tiribí collector (whose contributing sub-collectors are Damas and Cucubres).

12.3.4 Contamination from Agricultural and Industrial Activities

Costa Rica has an approximate annual consumption of 18 kilograms of pesticides per cultivated hectare and imports approximately 8,000 metric tons per year, of which 28 percent is considered to be of moderate to extremely toxic for human beings. Agriculture is highly dependant upon pesticides. The production of banana trees, one of the main economic activities of the Atlantic region, consumes approximately 40 percent of the total imported pesticides. As a result, most of the research focuses on analysing the effects of this type of contamination on health and the aquatic ecosystems (IRET 2000).

The main pesticides used in the region, with a high potential for water contamination, are herbicides, insecticides and fungicides. Studies of the basins of the San Juan, Tortuguero and Reventazón-Parismina Rivers have reported the presence of toxic substances used in the cultivation of banana trees. They can now be detected even in marine organisms found in the Cahuita National Park.

Table 12.3 presents some of the results obtained by IRET/UNA, based on the type of polluting agents and their effects on health.

The fungicides, insecticides and herbicides that are commonly used in agriculture, and in industrial wastes contain considerable amounts of heavy metals such as arsenic, thallium, cadmium, lead and mercury. These metals are polluting agents that are deposited in surface and groundwater sources and have harmful effects. Such metals can be found in the effluents from the manufacturing processes used for glass, silicone microchips, refined and welded metals, ceramics, textiles, paintings, wood, foods, medicines, and others. Their effects on humans can be acute especially from high levels of exposure, or ingestion in large doses. They can also be chronic, usually as a consequence of continuous exposure due to small repetitive doses over a long period. Both types of effects are summarised in Table 12.4, and are also analysed in the section on quality of water and public health.

Table 12.3. Contaminants found in the surface waters of the Atlantic region

River	Type of contamination	Consequences to health and aquatic ecosystems
San Juan, San Carlos and Sarapiqui Rivers	Clorotalonil, Cadusafos and Clorpirifos	Toxic fish Migraines, blurred vision, convulsions, fetal damage, renal damage.
Suerre River	Tiabendazol, Propiconazol, Imazalil, Terbuteros, Cadusafos, Clorpirifos, Propiconazol, Cadusafos, Carbofuran, Etoprofos	Nausea, vomiting, diarrhoea and death
Creeks of the banana zones	Organoclorate, organophosphate, Paraquat, Clorotalonil	Migraines, nausea, diarrhoea, convulsions, loss of consciousness

Source: IRET, 2000.

Table 12.4. Effects of heavy metals and agrochemicals on health

Acute	Chronic
Nausea	Renal insufficiency
Vomiting	Hepatic insufficiency
Tachycardia	Neumonitis
Diarrhoea	Neuropathies
Migraines	Loss of memory
Disnea	Hair loss
Convulsions	Cancer – lung, prostate, stomach
Hyper tension	Changes to the Central Nervous System (CNS)
Stomach pain	
Mialgias	

Source: MINSA, 2003.

The contamination of water via specific sources of liquid wastes is one of the main problems the country currently faces. As a solution, the establishment of an environmental measure of discharge under the polluters-pays-principle has been suggested. Those generating the contamination would be charged on the basis of the damages caused by the waste discharges affecting third parties as well as eco-systems.

The idea behind this proposal is to internalise the environmental costs generated by discharges, encourage industries to use cleaner production techniques and minimise the amount and concentration of pollutants being discharged into the water. It is also an attempt to generate income for the environmental authority in order to finance the prevention and control of water contamination. Those that surpass the permitted levels of chemical oxygen demand (COD) and total suspended solids (TSS) would be charged a specified amount (PROSIGA, CCAD, MINAE 2003).

Industrial, agro-industrial and agricultural activities are the main consumers of water as well as the greatest generators of contamination because of the effluents generated from their manufacturing processes. The Costa Rican Chamber of Industry, whose affiliates include more than 1000 companies of all sizes, is focusing on the issue through the National Centre of Cleaner Production, with the support of other organisations such as Centre of Technology Management and Industrial Computer Science (CEGESTI) and the Costa Rican Technological Institute (ITCR).

In recent years, the Chamber has provided incentives to affiliates who implement environmental management and cleaner production programmes. The aim is to promote systems that are more efficient in the use of water, energy and the wastes that result from the manufacturing processes, as well as ISO 14,000 certification. Implementation of these measures would generate savings that would increase benefits to the company in the short term. For example, in 2003, a tyre-producer implemented a programme for cleaner production internally which generated $500,000 in savings. By the end of the first year of implementation, the company had reduced by 50 percent its consumption of water and the amount of solid wastes, while its consumption of energy and solvents was cut by 15 percent (Reyes et al. 2003).

12.3.5 Water Quality and Public Health

Potable water is an indispensable requirement for ensuring good health of the population. Causes of the increase in the incidence of some of the water-borne diseases are directly related to the low coverage provided by the sanitary sewage system network (only 4.5 percent of the population of the country is served) and deficient potable water distribution (79.5 percent). In this area, MINSA, through the Department for the Protection of the Human Environment (DPAH), creates the tools needed for the provision of basic services, waste management, water and air contamination, discharge management, and others.

MINSA is responsible for overseeing monitoring and control to ensure enforcement of the legislation with the aim of protecting the population's health and the environment in general (Moreira 2004, personal communication). According to MINSA, the control of environmental problems does not depend on the adoption of policies and technical actions alone, but also on the provision of space for public participation, technical support and capacity building so that the different actors take part in the promotion and protection of environmental health. It should be noted, however, that although the DPAH has defined the roles to be played by both State and society in promoting health, in practice it has not been possible to work together to improve quality and offer potable water to 100 percent of the population.

While legally MINSA is mandated as the entity in charge of promoting water quality, in practice its role is practically invisible. According to AyA, the main causes for this inefficiency are lack of trained personnel, insufficient budget, and above all absence of water pricing. Currently, water quality monitoring is delegated to each of the health departments for the nine defined geographical regions. Along with their regular functions, these departments must also carry out sanitary inspections and educate the population about water.

Instruments do exist, in the shape of sanitary bylaws and norms for water supply systems, that require water of potable quality from the operators, but they have never been implemented. Due to inefficiency in the application of the instruments, MINSA has initiated a process of strategic change. The Ministry has begun informing civil society on the quality of water for human consumption and the health risks implied. The purpose is to encourage civil society itself to demand the corresponding improvements from the different entities operating the water supply systems. Nevertheless, this is a process that has just begun and results are still not evident. Costa Rican society is highly passive with respect to water quality and has delegated all responsibility to the State and its organisations.

The fact that more than 20 percent of the population does not receive potable water has serious implications in terms of health. The increase in the incidence of water-borne diseases due to the consumption of non-potable water, and increase in infectious agents, demonstrates that the country is on the verge of a health crisis.

Already diarrhoea is surpassed only by AIDS as the leading cause of death from diseases for which declarations are obligatory.

A summary of such diseases most frequently associated with water is provided in Table 12.5. From this Table, it can be seen that, in 2002, around 4 percent of the population suffered from diseases transmitted by water and/or related to water. The proportion, however, could in fact be much greater. Not all people suffering from illnesses consult health centres, and cases of patients who only go for private consultation or to pharmacies go unreported.

Table 12.5. Obligatory declarations of diseases related to water, Costa Rica, 2002

Disease	Carrying agent	No. of cases
Diarrhoea	Shighela, E.Coli Virus	138,410
Leptospirosis	Leptospira	298
Hepatitis A	Hepatitis A Virus	230

Source: MINSA, 2003.

Between 1999 and 2003, 12 major incidences of diarrhoea were identified in which the water for human consumption was the transmitter of the different agents (virus, bacteria). In all of these cases, the analyses done by the Costa Rican Institute of Research and Teaching in Health and Nutrition (INCIENSA) on the water for human consumption detected the causal agent of the disease in both water and faeces of the patients.

Such incidences emphasise the reasons why the country must make proper water quality management a priority issue. As well as its toll in human suffering, diarrhoea exacts a major social and economic cost. The diagnoses of the 138,410 cases of diarrhoea reported by the health centres of the country during 2002 meant an approximated cost of $31 million to the Costa Rican Social Security (CCSS) system. This amount is underestimated in that it covers only the costs of attending patients in peripheral clinics or EBAIS. It does not include an estimation of the cases that go unreported, or the medicines, laboratory examinations and, where necessary, stays in specialised hospitals.

Children and older people are particularly susceptible. In the case of children, repetitive bouts of diarrhoea (as in the case of Santa Barbara of Heredia, Dota and Talamanca) can lead to undernourishment, failure to progress, scholastic underperformance and severe susceptibility to other infections and diseases.

The National Water Laboratory, through its Sanitary Risk Programme, has identified and demonstrated the risks posed by the municipal water supply systems involved in the cases of diarrhoea between 1999 and 2003. Of the 27 systems that were studied, 92 percent were found to be of intermediate to very high risk.

Continuous monitoring and maintenance of the sanitary systems are necessary in order to reduce vulnerability to water-borne diseases. When necessary improvements are not made to the system, cases such as those of communities like Santa Barbara of Heredia, Santa Maria de Dota and Talamanca appear where incidences of diarrhoea are repetitive. Furthermore, diarrhoea is one of the most frequent reasons for consultations in CCSS and EBAIS health clinics.

Because of the dynamics of the supply of basic services, Costa Ricans assume that they are the sole responsibility of the State. In the particular case of water, the service has been practically free, with the involvement of citizens in many cases limited to the construction of infrastructure and later to making up the board of directors. Consequently, the Costa Rican population lacks a proper appreciation of the true value of water.

12.3.6 Investment in Water Quality

The largest investments in infrastructure for sanitation and water for human consumption were made during the 1960s and 1970s, as a product of a model of economic development based on import substitution or inward development. The era was characterized by a beneficent State, promoting investments in education, health and infrastructure while supporting policies that protected the nation's producers. During this period, the State was regarded as the sole entity responsible for investing in water for human consumption and sanitation. Water was regarded as an abundant and free resource.

A combination of domestic factors and external pressures provoked by the debt crisis of the 1970s and early 1980s produced a change towards a more outward-looking development relying on structural adjustment programmes. Policies of cost reduction, market liberalisation and export promotion were applied, with the State moving away from its role as benefactor and investor and beginning to assume that of an accountant and regulator. During the 1980s and the 1990s, budget restrictions affected infrastructural investments, as well as spending on the monitoring and maintenance of the sanitary systems and water for human consumption, and preventive medicine.

Table 12.6. The impacts of the Investment Plan of Tariffs

Area	No. of connections 2000 (thousands)	No. of connec-tions 2020 (thousands)	Total investment 2001-2020 (thou-sand/$)	Equivalent monthly cost per connec-tion ($/month)	Present tar-iff by con-nection ($/month)	Pre-dicted increase (%)
Urban Potable water	520	964	722,000	4.45	9.10	49
Sewage system	179	870	684,000	6.75	2.74	24
Rural Potable water	261	332	120,000	1.75	3.66	48
Sanitation			69,000			

Source: AyA, OPS, OMS, 2002.

Between 1991 and 1998, State spending averaged $20 million a year. As a result, it is estimated that annual investment of $80 million (AyA, OPS, OMS 2002)

is now needed to cover the deficit. As indicated in Table 12.6, the effect of the increase in investment, if transferred entirely to the users, would mean an increase of 247 percent in the present tariff of the sanitary sewage system for urban areas, and an increase of 49 percent in the potable water surcharge for the countryside.

Of the total budget for 2002, the Health Ministry was allocated 2.1 percent ($77.9 million), of which 2 percent was for monitoring and control of the human environment. This represents 0.01 percent of GDP, as can be seen in Table 12.7. The specific functions for which these funds are earmarked are for the protection of the human environment through monitoring, research and control of the risk factors and of the physical conditions of the facilities. The aim is to provide a healthy environment for companies, workers and the population in general.

Table 12.7. Budgetory allocations for water quality according to the total budget of the country and Gross Domestic Product

Budget category	2002 (million dollars)	Relation investment- total Budget	Relation investment- GDP
Ministry of Health	77.9	2.11	0.46
Monitoring and control of the Human Environment	1.5	0.04	0.01
Costa Rican Institute of Water Supply and Sewage Systems	74.3	2.02	0.44
Direct Investment in Urban Works[1]	6.1	0.16	0.04
Spending according to function:	77.9	2.12	0.46
Health	73.3	1.99	0.44
Environment Sanitation	0.1	----	----
Potable Water and sewage system	4.5	0.12	0.03
Government of the Republic of Costa Rica	3,683	100.00	21.90
Gross Domestic product	16,818	----	100.00

Source: General Accounting Department of the Republic, 2003.
Notes: [1]Investments in improvements, extensions and construction of water supply systems, tanks and treatment plants.
Exchange rate of US dollars to colon: 359.82.

The Institutional Memorandum and the spending report presented to the General Accounting Department of the Republic (2003) by AyA, this organisation invested $6.1 million in urban works. This represented 8.2 percent of AyA's total budget, and 0.04 percent of that of the country. Some 85 percent of these funds came from the sale of potable water and sewage system services. The funds were invested in the construction of treatment plants, tanks, improvements and refurbishment of existing water supply systems, as well as the extension of water supply in the metropolitan areas of San José and several rural areas.

Some 2.12 percent ($77.9 million) of the national budget and 0.46 percent of GDP were dedicated to hygiene, environmental sanitation, sewage systems and potable water, in line with the budget assigned to the Health Ministry. Investment in monitoring and control of the human environment, sewage systems and sanitation amounted to 0.05 percent of GDP. Nevertheless, AyA estimates that the investment necessary to cover the deficiencies of the sanitary sewage system at the national level averages $157 million, which is 32 times the present level.

In 2002, the country invested $7.6 million in monitoring, sewage systems and sanitation, four times less than what was spent on treating cases of diarrhoea ($31 million). This demonstrates a clear tendency, at least at the policy level, to invest in the cure rather than in prevention.

On the other hand, payments for environmental services amount to 42.4 percent of the budget of the Ministry of Environment and Energy; 0.26 percent of the total budget; 0.06 percent of the GDP; and twice as much as the investment in the sewage systems and for potable water. This demonstrates a clear policy aimed at the protection and conservation of water services rather than sanitation and disease prevention.

In addition, investment in potable water and sanitation is compared with the amounts for other sectors, it corresponds to 0.75 percent of the budget for education (4.8 percent of GDP) and 8.35 percent of spending on health

12.4 Conclusions

- In Central America, the water regime is characterised by the absence of clear policies, outdated or inadequate legislations, and overlapping functions between governmental, private or external governing bodies, supervisors and executors, thus complicating the administration of water resources and decision-making at the political level.
- The region lacks a vision of integrated management of water resources, thus favouring sectoral policies. Investments in the protection and conservation of water-related services have been given priority, pushing issues surrounding potable water and sanitation into the background. Costa Rica, Guatemala and El Salvador, for example, negotiated loans with the World Bank to strengthen or implement national programmes of payment for environmental services.
- Domestic wastewater discharges are the main sources of water contamination because they mostly go untreated. Costa Rica is a clear example: only about 25 percent of the population is served by the sanitary sewage system and only 2.4 percent is served by both the sanitary sewage system and treatment plants.
- Sanitation systems in the region have collapsed, leaving most of the population without adequate services. In the case of Guatemala, however, almost 95 percent of the population has access to the sewage system, but almost 100 percent of waste discharges flow into the rivers without any treatment. An investment of $1.58 billion dollars is required to cover the current needs.
- At best, investment in infrastructure and maintenance within the potable water and sanitation sector is still only enough for minimal maintenance for operational needs. Costa Rica, for example, dedicates only 0.04 percent of GDP to that end.
- It has been said that the system of water supply for human consumption in Central America covers most of the population. Nevertheless, requirements with

respect to quality are not being met. An investment of $1.48 billion (BID 2003) would be needed to provide most of the population with potable water and appropriate distribution systems that reduce losses and contamination as a result of obsolete networks.

- The total investment required to meet demands for potable water and sanitation in Central America is $3.06 billion (BID 2003), or about 5.8 percent of the GDP of the region. Current investments amount to no more than 1 percent of GDP. In order to cover the deficit, funds must be redirected from other sectors or activities, a move that would require strong political wills from the regional governments.

Additional factors that detract from the importance given by the governments to problems associated with potable water and sanitation are:

1. Lack of awareness among decision-makers of the impacts of lack of investment in water and sanitation on health and poverty alleviation;
2. Failure to plan for opportune investments in infrastructure and maintenance of the distribution systems and water treatment;
3. Lack of effective regulations for the protection and control of water sources;
4. Centralisation of management in capital cities;
5. Lack of financial resources and equipment for monitoring, for example, one laboratory for the entire country in the case of Costa Rica; and
6. Shortage of qualified and trained personnel.

- Figures shown by official sources, OPS-OMS, and State of the Region, on coverage and distribution of water for human consumption and environmental sanitation are underestimated if analysed on the basis of incidence of water-borne diseases, such as diarrhoea, that have a high frequency. Indices in Costa Rica, Panama and Belize are lower than those of the rest of the countries.
- The lack of appropriate investment in infrastructure and maintenance, along with other previously indicated factors, have resulted in the proliferation of water-borne diseases such as diarrhoea and hepatitis throughout the region. In Costa Rica, it is estimated that the State invested nearly $31 million on the control of diarrhoea alone in 2002.
- Through methods of economic evaluation (e.g. costs of diseases), estimates of external costs associated with water-borne diseases indicate that the medium- and long-term investments made in disease control would be higher than the investment at current prices in infrastructure and maintenance of appropriate systems of water treatment and sanitary sewage systems.
- It is therefore urgent, at regional level, to implement a preventive rather than remedial policy for public health. This would allow governments to avoid damages before they occur and reduce the incidence of water-borne diseases through an integrated vision, reflected in policy planning.
- In almost all the countries of the region, environmental conscience at the public health level is scarce, or even non-existent. What prevails is a culture based on a paternalistic vision, in which the State is the sole entity in charge of ensuring the quality of the potable water and sewage systems. As revealed by

the Costa Rica case study, few community initiatives exist for the prevention of water-borne diseases. Nevertheless, populations have shown a greater sensibility towards "caring for" the water supply, thus in a way changing the vision of water as a free resource to which every one has free access.

- One aspect seldom discussed in terms of water quality involves the problems of contamination from industry and farming. Most of the countries of the region lack appropriate systems for monitoring or regulations that could force a reduction in the use of agrochemicals in certain crops, or the treatment of the liquid or solid wastes generated by industry.
- It is important to recognize that efforts are now being made in the private sector to reduce the amount of water used and discharges produced from industrial processes. Examples include the programmes for cleaner production and environmental management that provide incentives through chambers of industry at regional levels.
- The current trends are towards poorer water quality within the region. The countries of Central America lack the capacity to take on the challenge of investment in the short term. Many do not even have the regulations, plans, or the political urgency that would allow for short-term solutions. What matters is making improvements not only to the physical and chemical characteristics of water, but also to the social and economic conditions of the population.

12.5 References

Araya D, Araya A, Trejos S (2003) Estudio sobre la situación de la Tecnología de Tratamiento de las Aguas Residuales de Tipo Ordinario en Costa Rica. Organización Panamericana de la Salud, Organización Mundial de la Salud, Instituto Costarricense de Acueductos y Alcantarillados, San José, Costa Rica

Araya D (2000) Pequeñas plantas de tratamiento y estaciones de bombeo de aguas residuales. Área Metropolitana. Operaciones de Sistemas y Aguas Residuales. AyA, San José, Costa Rica

Araya D (2001) Polo Turístico Papagayo. Plantas de tratamiento de aguas residuales. Instituto Costarricense de Acueductos y Alcantarillados (AyA), San José, Costa Rica

AyA (1998) Estudio de Alcantarillado Sanitario de la Gran Area Metropolitana. Informe Final. Instituto Costarricense de Acueductos y Alcantarillados. Geotécnica. Volumen 1, San José, Costa Rica

AyA, OPS, OMS (2002) Agua potable y saneamiento de Costa Rica: análisis sectorial. Instituto Costarricense de Acueductos y Alcantarillados, Organización Panamericana de la Salud y Organización Mundial de la Salud, San José, Costa Rica

BID (2003) Las Metas del milenio y las necesidades de inversión en América Latina y el Caribe. Conferencia Internacional, financiación de los servicios de agua y saneamiento: opciones y condiciones, 10-11 Noviembre, Washington, DC

BCCR (2004) Central Bank of Costa Rica. http://www.bccr.fi.cr

Céspedes M (1993) Evaluación de la operación y mantenimiento de las Lagunas de Estabilización de Liberia Guanacaste. Tesis para el grado de Licenciatura en Ingeniería Civil, Universidad de Costa Rica, San José, Costa Rica

Davis R, Hirji R (2003) Water resources and environment. Technical Note No. 1: Water quality assessment and protection. World Bank, Washington, DC

Fallas (2004) Personal Communication. Ministry of Health, San José, Costa Rica

FAO (2003) Sistema de Información sobre el Uso del Agua en la Agricultura y el Medio Rural. Organización de las Naciones Unidas para la Agricultura y la Alimentación, Roma, Italia

General Accounting Department of the Republic (2003) Institutional Memorandum 2002, San José, Costa Rica

INEC (2000) Censo de Población 2000. Instituto Nacional de Estadísticas y Censos, San José

IRET (2000) Reducción del Escurrimiento de Plaguicidas al Mar Caribe. Informe Nacional: Costa Rica. Instituto Regional de Estudios de Sustancias Tóxicas. Universidad Nacional. Heredia, Costa Rica

La Nación, 24 de enero 2004, p 6A, San José, Costa Rica

MINSA (2003) Boletín Epidemiológico. Sistema Nacional de Vigilancia de la Salud. Ministerio de Salud, San José, Costa Rica

Mora D, Portuguéz F (2000) Diagnóstico de la cobertura de agua para consumo humano en Costa Rica. San José, Costa Rica

Mora D, Portuguéz F. (2002) Diagnóstico de cobertura y calidad del agua para consumo humano. Revista Costarricense de Salud Pública 9(16): 9-15, San José, Costa Rica

Moreira (2004) Personal Communication. Ministry of Health, San José, Costa Rica

LNA (2003) Vigilancia de la calidad del agua suministrada por acueductos operados por ASADAS. Laboratorio Nacional de Aguas, Instituto Costarricense de Acueductos y Alcantarillados, Tres Ríos, Costa Rica

OPS , OMS (2000a) Evaluación de los sistemas de agua potable y saneamiento 2000 en las América. Informe analítico de Belice. Organización Panamericana de la Salud y Organización Mundial de la Salud. http://www.cepis.ops-oms.org/eswww/eva2000/ belice/informe/inf-04.htm

OPS, OMS (2000b) Evaluación de los sistemas de agua potable y saneamiento 2000 en las América. Informe analítico de Guatemala. Organización Panamericana de la Salud y Organización Mundial de la Salud. http://www.cepis.ops-oms.org/eswww/eva2000/ Guatemala/informe/inf-04.htm

OPS, OMS (2000c) Evaluación de los sistemas de agua potable y saneamiento 2000 en las América. Informe analítico de Honduras. Organización Panamericana de la Salud y Organización Mundial de la Salud. http://www.cepis.ops-oms.org/eswww/eva2000/ Honduras/informe/inf-01.htm

OPS, OMS (2000d) Evaluación de los sistemas de agua potable y saneamiento 2000 en las América. Informe analítico de El Salvador. Organización Panamericana de la Salud y Organización Mundial de la Salud. http://www.cepis.ops-oms.org/eswww/eva2000/ salvador/informe/inf-04.htm

OPS, OMS (2000e) Evaluación de los sistemas de agua potable y saneamiento 2000 en las América. Informe analítico de Nicaragua. Organización Panamericana de la Salud y Organización Mundial de la Salud. http://www.cepis.ops-oms.org/eswww/eva2000/ Nicaragua /informe/inf-01.htm

OPS , OMS (2000f) Evaluación de los sistemas de agua potable y saneamiento 2000 en las América. Informe analítico de Panamá. Organización Panamericana de la Salud y Organización Mundial de la Salud. http://www.cepis.ops-oms.org/eswww/eva2000/ /Panama /informe/inf-04.htm

OPS, OMS (2001) Informe Regional sobre la Evaluación 2000 en la región de las Américas: agua potable y saneamiento, estado actual y perspectivas Organización Panamericana de la Salud, Washington DC

PNUD (2003) Agua para todos, agua para la vida. Informe de las Naciones Unidas sobre el Desarrollo de los Recursos Hídricos en el Mundo. Programa de las Naciones Unidas para el Desarrollo, New York

PROSIGA, CCAD, MINAE (2003) Canon Ambiental por Vertidos: Un instrumento económico para el control y la prevención de la contaminación hídrica en Costa Rica. Programa Mesoamericano de Sistemas de Manejo Ambiental, Comisión Centroamericana de Ambiente y Desarrollo, y Ministerio de Ambiente y Energía, San José, Costa Rica

Proyecto Estado de la Nación en Desarrollo Humano Sostenible (2002) Estado de la Nación en Desarrollo Humano Sostenible. Noveno Informe 2002, San José, Costa Rica

Reyes V, Segura O, Gámez L (2003) Régimen del Recurso Hídrico: El caso de Costa Rica. GWP-Centroamérica-CINPE-UNA, San José, Costa Rica

Reynolds J (2002) Manejo Integrado de Aguas Subterráneas: un reto para el futuro. Editorial Universidad Estatal a Distancia, San José, Costa Rica

UNEP (2003) GEO Latin American and the Caribbean: Environment Outlook 2003. United Nations Environment Programme, Mexico

Legal and Institutional Frameworks for Water Quality Management in Argentina

Lilian del Castillo

13.1. Introduction

Environmental regulations aim to protect nature, minimise harmful modifications of a human origin and correct the negative impacts caused by environmental degradation. Regulations for environmental protection aim to protect people from their own behaviour, and environmental protection focuses on conserving living conditions in their broadest sense, including human quality of life and the biota, which are closely interlinked, while trying to achieve the sustainable use and management of natural resources. Sustainability is an element of environmental protection: it is a conceptual framework that gives direction to both the content and purpose of environmental protection.

The possibility of using water and other natural resources, and even the right to use them, has never been an issue. But the concept of use needs to be regulated in order to avoid regular misuse, or at least to reduce it, and especially in order to prevent use becoming abuse.

The legal environmental protection regime requires different actions for every natural resource. This is especially so for such an essential element as water, as different needs have to be addressed and a large number of specific situations have to be considered. Environmental protection regulations apply to both water quantity and quality, although this chapter takes into account only those aimed at the conservation of water. This is because water is directly connected to the applicable regime and depends almost totally on regulatory standards, or rather to their absence.

Protecting water quality means facing the same difficulty as found in other areas of environmental protection. Problems are difficult to solve as people are often unaware that there is a problem. This is why, due to the extent and range of the subject, it is hard to describe the legal and institutional regime that regulates the different aspects of water quality in Argentina. In this chapter, two basins with serious pollution problems and two others that have been successful in establishing management regimes, will be discussed.

Water quality is a subject in which human influences that directly affect the resource can be regulated. Having a wide variety of uses can be the most important cause of deterioration but regulation can be used to rectify this situation. The use of different parameters proves the need to incorporate regulations and to check if these, or their application, are appropriate.

Different natural and human factors have an impact on water in its natural state. When the impact is negative, is considered it as pollution. The purpose of the regulation aimed at preserving or restoring certain water quality is known as environmental protection, which aims to preserve the resource for human use and the preservation of aquatic life. The scientific basis of environmental protection is the interdependence of living beings.

Water quality regulations refer both to water for human consumption and to ambient water. However, while regulations on drinking water are the responsibility of public health institutions, regulations applied to ambient water are the responsibility of environmental authorities.

Legal and institutional multiplicity shows that a scarcity of regulations does not impact on the proper management of the environment, but a lack of systematisation does. As a corollary, the overlapping of responsibilities among the pertinent bodies and authorities helps to blur their responsibilities, a fact that significantly decreases their commitment to the results they should be trying to achieve. Consequently, to regulate the environment in an efficient manner, there should be a proper and tenacious institutional structure that meets the goal of preserving legally protected interests.

Adopting a certain water quality regulation is only the first step in protecting the resource. Its application will determine whether the criteria are proper and whether the adopted mechanism and methodology are efficient, as well as the right ones. This can be assessed objectively by comparing quality conditions before and after the introduction of any regulation.

Pollution control is the main objective of current legislations and of international agreements on preserving the quality of the aquatic environment through regulations that use different mechanisms to penalise polluting activities. An agreement dealing in part with this topic is the Treaty Concerning the Río de la Plata and the Corresponding Maritime Boundary (United Nations Treaty Series, Vol. 1295:294; Approved in Argentina by Law No. 20.645, Official Gazette) between Argentina and Uruguay, signed in Montevideo on 19 November 1973, which states that "each party undertakes to protect and preserve the aquatic environment and, in particular, to prevent its pollution by enacting appropriate rules and adopting appropriate measures in accordance with applicable international agreements and adjusted, where relevant, to the guidelines and recommendations of international technical bodies" (Section 48). Although the treaty is applicable only to those who signed it, it should be underlined the definition of pollution as an element to be construed as follows: "For the purposes of this treaty, pollution shall mean the direct or indirect introduction by man into the aquatic environment of substances or energy which have harmful effects" (Section 47).

Protection of the aquatic environment, with the purpose of protecting the biota, is a permanent objective and it has dynamic characteristics in the sense that it should be flexible in terms of methods and procedures, and not divided into rigid categories. The regulation that supports environmental protection should be limited to constructing the legal framework of rights and duties and the goals for environmental quality by creating guidelines that the bodies with proper jurisdiction can apply, depending on specific circumstances.

13.2 Environmental Rules in Argentina's Federal System

13.2.1 Different Levels of Water Quality Legislation

In a federal country like Argentina, legislation is fragmented, as reflected in the different applications of law, the variety of legal bodies and authorities, and the different forums where environmental problems can be discussed.

Since 1986, a number of provincial constitutions have incorporated regulations on environmental protection, i.e., Constitutions of the Province of Jujuy (1986), Section 22; of the Province of La Rioja (1986 and 1998, Section 66); of the Province of San Juan (1986, Section 58); of the Province of Salta (1986 and 1998, Section 30); of the Province of San Luis (1987, Section 47); of the Province of Río Negro (1988, Sections 84 – 85); of the Province of Tucumán (1991, Section 36); of the Province of Tierra del Fuego, Antártida e Islas del Atlántico Sur (1991, Sections 54 – 55). Moreover, other provinces incorporated provisions about environmental protection and the right to a healthy environment in accordance with the amendment of the National Constitution in 1994, such as, Constitutions of the Province of Chaco, Section 38; of the Province of Chubut, Sections 109 – 110 – 111; of the Province of La Pampa (1994, Section 18); of the Province of Santa Cruz (1998, Section 73); of the Province of Córdoba (2001, Section 11); and, of the Province of Santiago del Estero (2002, Section 35).

However, only with the 1994 constitutional amendment, the legal principle of environmental protection became a federal rule. The specific regulation, Section 41, addresses the scope of action that can be taken by the State and the rights of individuals concerning the environment (first paragraph). It also refers to the obligation of the pertinent bodies and authorities to protect individual rights (second paragraph). In addition, it states that the writing of an environmental legal framework comes under the national jurisdiction, while additional provincial regulations (third paragraph) can be added by provincial authorities.

The regulation states that people have a right to enjoy a healthy environment and are duty bound to preserve it because "all inhabitants have the right to a healthy and balanced environment that is suitable for human development. So that productive activities must be carried out without jeopardising the needs of future generations, and it is the duty of those who conduct productive activities to preserve the environment" (Section 41, first paragraph).

The right to a healthy environment means something different in the chapter on rights and guarantees of the Constitution, which considers the individual as an active and passive subject who can invoke it but who is also obliged to comply with it. An individual not only has the right to enjoy a healthy environment but also the obligation to protect it. The State has to establish the proper legal and institutional structure so that exercising one's acknowledged right is not infringed. The constitutional text states that "authorities shall provide for the protection of this right" (Section 41, National Constitution, second paragraph). The duty of the State is to guarantee the exercise of the acknowledged right, avoiding incidences of environmental degradation and taking all necessary measures to avoid it.

The constitutional regulation also acknowledges that human activities degrade the environment and that the law should provide the tools to modify such harmful aspects (Valls 1999:36-40). Individuals have the legal right to protect themselves from the activities of other people within a legal framework in which to exercise this right (Gambier and Lago 1995). Undoubtedly, whatever the National Constitution says about the violation of civil rights or guarantees is applicable. It states that "any person can immediately bring and expedite a case whenever his/her civil rights or guarantees have been violated, provided there is no other suitable legal means available to take against the public authorities or private individuals who, by neglect or action, have injured, restricted, altered or threatened, with evident arbitrariness or illegality, the rights and guarantees acknowledged by this Constitution, a Treaty or a Law" (National Constitution, Section 43, first paragraph). It expressly mentions "all the rights that protect the environment" (National Constitution, Section 43, second paragraph).

The scope for taking action in a case where civil rights or guarantees have been violated is much greater due to the fact that the derogated statute 17.45 limited its action to the violation or ignorance of constitutional regulations, even though its rules could be applied in actions brought by state authorities or private individuals. Current law not only legitimises the actions of concerned individuals but those of the ombudsman and environmental associations as well (Section 43, second paragraph). Thus, the constitution has made progress on the permanent underlying threat of turning granted rights into abstract promises and has created a procedural system for ensuring its application.

Constitutional rules incorporate rights that can be invoked not only before state bodies but also before other individuals, corporations or organisations of different kinds, and it simultaneously strengthens control systems. The application of the rights and duties listed in the constitutional provision implies adopting a new set of regulations where principles are incorporated and regulated, thus linking elements pertaining to public law and private law. Regulatory identity is grouped in a specific branch of law, called environmental law.

In countries with a federal structure, such as Argentina or the United States of America, whose Constitution was a model for the Argentine Constitution, Germany or Brazil, regulations that cover environmental protection and application procedures at a federal level have to be included in a national/provincial integrated management of the environment in general, and water resources in particular. Federal legislation is essential in order to elaborate a comprehensive environmental policy and it is vital that basic regulations be uniform and compatible so that national policy on the environment can be established and put into practice. Therefore, there are federal entities with jurisdiction to set the general standards for environmental protection as well as control systems to supervise their implementation.

Together with the national authorities, local bodies exercise their own jurisdiction in order to apply the national environmental policy, using additional regulations and their own monitoring and control procedures that are adapted to the specific situation of each provincial state. As a consequence, according to the constitution, the State issues and applies the minimum premises relating to environmental protection while provincial states issue and apply the additional local legislation within those premises. In this way, federal and provincial jurisdiction is applied to environmental protection. This is not to imply a failure to accept municipal jurisdiction on environmental protection. As a matter of fact, municipal governments recognise at a constitutional level (National Constitution, Section 5) that, although they must adjust their regulatory guidelines to national and provincial regulations, they are autonomous as regards the approval and application of regulations on environmental protection, authorisation for the installation of stores and industries, use of water, dumping of wastes in the water bodies, and so on.

Regarding national jurisdiction on the issue of environmental protection regulations, the Constitution states that "the nation shall regulate the minimum protection standards and the provinces those necessary to reinforce them, without affecting their local jurisdiction" (Section 41, paragraph 3). The nation's legislative body is the National Congress, which is in charge of enacting regulations that establish basic criteria applicable across Argentina. The provinces adopt local regulations that are compatible with national ones. The National Congress is empowered to legislate on substantive matters.

13.2.2 Concurrent Water Quality Jurisdiction

Setting jurisdiction limits causes conflict between the Nation and the provinces. This conflict has worsened since the 1994 constitutional amendment because of the confusion it has caused regarding the ownership of natural resources.

On this issue, Section 124, second paragraph, states that "provinces own the natural resources existing in their territory." Some subsequent provincial constitutions seem to state that this qualification has been construed as a synonym of exclusive provincial public ownership.

The provinces do not support national legislation that states the minimum standards for environmental protection, according to what is set in Section 41 of the Constitution. However, only through national legislation that has a national scope can efficient water quality management be achieved. Minimum environmental standards must not just be abstract statements, but they should lead to the development of principles and parameters and must be based on institutional mechanisms of sampling, assessment and performance, carried out through the appropriate bodies and authorities. Local authorities can complement the application of national legislations for environmental protection by applying other parameters and creating their own pertinent bodies and authorities.

In federal states, such as Brazil, the national government has the power to adopt water policies because it has wider jurisdiction than provincial authorities. In Argentina, opposition to national jurisdiction creates an institutional and regulatory crisis that makes it impossible to adopt an efficient policy on water quality management.

The enacted federal legislation establishes the minimal environmental norms for water preservation. Statute 25.688 of November 28th, 2002 (Official Gazette, January 3rd, 2003) states that pertinent bodies and authorities are in charge of setting the environmental parameters and standards for water quality and of designing the National Plan for its preservation, exploitation and rational use. The purpose is to have a legal framework that makes it possible to adopt a policy on the environmental management of water that can be applied across Argentina and complemented by local authorities, depending on the characteristics of the water resources and the needs of the community in each region.

As for the structure of the National Public Administration, the Ministry of Public Health should adopt all necessary regulations to facilitate compliance with the constitutional objective of preservation and protection of the environment, carrying out a policy that aims to reach the goal of sustainable development (Decree 141/2003, Section 23, paragraph 41). As a consequence, it is the appropriate national body or authority that sets regulations for environmental protection, including water.

The Secretariat of the Environment and Sustainable Development has been part of the Ministry of Public Health since 2003. It is the body responsible for boosting "the environmentally sustainable management of water resources, in coordination with the Ministry of Federal Planning, Public Investment and Utilities." It is also in charge of applying international conventions on environmental protection and on uses and water quality that Argentina has signed (Decree 141/2003, Section 23, paragraphs 46 and 54; Decree 295/2003, paragraphs 16 and 14), such as the Basle Convention Agreement on the transportation of hazardous wastes (Convention on control of cross-border movements of hazardous wastes and their disposal, Basle, adopted in 1989 under the management of UNEP, approved by Argentina by Statute 23.922, March 21, 1991), and the Ramsar Convention (Convention related to wetlands of international importance, especially the habitats of water birds, Ramsar, Iran, 1971, approved by Argentina by Statute 23.919, March 21, 1991).

The Under-secretariat of Water Resources, which is a part of the Secretariat of Public Works from the Ministry of Federal Planning, Public Investment and Utilities, has jurisdiction over "the elaboration and performance of the national water policy." It is in charge of "suggesting the regulatory framework for the management of water resources, linking and coordinating the action of the other jurisdictions and bodies that are governed by the water policy." The under-secretariat's responsibility covers both domestic and the international water resources shared with neighbouring countries.

It is in charge of "formulating and implementing management and infrastructure development programmes and actions and of services linked to water resources relating to their construction, operation, maintenance, control and regulation at international, national, regional, provincial and municipal levels, as well as implementing participation mechanisms for the private sector and the community, as the case may be" (Decree 1283/2003 and Decree 27/2003, May 27th, 2003, Ministry of Federal Planning, Public Investment and Utilities functions). Both entities share the sustainable environmental management of water resources nationally. The Ministry of Public Health has specific jurisdiction over water quality, and the Under-secretariat of Water Resources is responsible for hydrological aspects, infrastructure projects and water policy.

13.3 Water Quality Rules in Argentina

13.3.1 General Rules

The principle of sustainability has changed the concept upon which the activities that have a negative impact on the environment were legislated. This principle, together with others that are necessary for an environmentally sustainable use of waters, have been incorporated into Argentine legislation.

In 2002, the General Environmental Law was enacted and definitions and objectives that will make it possible to create a national environmental policy were adopted (Statute 25.675, of November 6, 2002). The new law describes legally protected interests and lays out the following as principles of environmental policy: congruence, prevention, precaution, intergenerational equity, progress, responsibility, subsidies, solidarity and cooperation (Section 4). The law defines "minimum protection standards" as any regulation "that grants uniform or common environmental protection for the whole national territory." It states that minimum standards must "guarantee the dynamics of ecological ecosystems, keep their load capacity and, in general, ensure environmental preservation and sustainable development" (Section 6). This description seems contradictory in itself, due to the fact that keeping the load capacity implies permitting the polluting activity. It is also inconsistent with regard to the principles of prevention, precaution and intergenerational equity.

In fact, according to the definition of sustainable development in the Bruntland Report (1987), 'Sustainable development is development that meets the needs of the present without compromising the ability of future generations to meet their own needs' (Our Common Future 1987:43) and the limitations and precautions that it implicitly includes (Tortajada 2001:6-14), the starting point of pollution control should not be the self correction capacity of the recipient body, but the environmental protection objective to preserve the resource and protect both the biota and the quality of life. Pollutants and other harmful elements must be treated and disposed of only when they are harmful for the water body.

In 1991, the first regulation on national pollution control was enacted. It was limited to the generation, manipulation, transportation and treatment of hazardous wastes (Statute 24.051 of December 17, 1991, and Decree 831/1993 of April 23, 1993). The regulation is applied in the metropolitan area of the city of Buenos Aires and other regions under national jurisdiction. In the provinces it is only applicable if wastes are carried from one province to another or can cause harm outside their province of origin. These provisions mainly follow the structure of the Basle

Convention on the transportation of hazardous wastes, in force since May 5th, 1992.

Nevertheless, national legislation is not limited to the movement of hazardous wastes across national borders but simultaneously regulates the generation, treatment and final disposal of such wastes. International regulations on the cross-border transportation of hazardous wastes are not regulations meant to control pollution, but an addition to internal legislation that does not regulate waste disposal beyond national borders.

In Argentina regulations on pollution control and cross-border transportation of hazardous wastes were incorporated in the same legal instrument, thus resolving the legislative gap that was inconsistent with the approval of the Basle Convention. Additional regulations defined the concept of "activities that generate hazardous wastes" (Decree 831/93) and included other elements such as hospital clothes (Resolution 221/2000 of the Ministry of Social Development and Environment, November 4, 2000) due to the fact that governmental agencies have the competence to enlarge the list of hazardous wastes (Resolution 897/2002, August 23, 2002).

In 2002, other regulations were issued on pollution control, and they referred to the same minimum protection standards for the environmental management of industrial wastes and their comprehensive management (Statute 25.670 of October 23, 2002, partially suspended by Decree 1343/2002). Therefore, the subject was extended to all types of wastes, not only hazardous, and was applied across the whole country. The new law established civic responsibilities and also expanded the regime of responsibilities, infringements and penalties for the whole country, but it did not appoint the appropriate bodies and authorities. The minimum standards for the management and disposal of PCBs (polychloride bisphenol) were also approved.

Regarding the specific matter of protecting the quality of both surface and groundwater, Decree number 674/1989 was established in the metropolitan area of the city of Buenos Aires (Decree 674/1989 of May 24, 1989, amended by Decree 776/1992 of May 12, 1992). It includes waste or sludge being dumped into the sewerage system and/or drainpipes or watercourses at industrial and other premises.

The control system adopted is in line with the polluter-pay-principle, by which the users that do not comply with the set parameters must pay a special charge for pollution control calculated according to the daily volume of effluent, and which will increase every year for 20 years, starting from 1998 (Section 7). A company dumping forbidden wastes can face penalties ranging from fines to the closure of its drains. Said penalties can last until the required work to improve the quality of the wastes is carried out.

If the company shows that it is taking measures to comply with the permissible limits, it can be exempt from the payment of stamp duties (Section 6). The regulation also bans the accumulation of solid wastes that may represent a pollution hazard in surface and groundwaters (Section 12). The values of permissible limits and those transitorily tolerated shall be adopted by the appropriate body or authority "on the grounds of the step set for the quality guideline values of the water course" (Section 5). The implementing authority was the state-owned utility that provided water and sanitation, National Water Works. This operation was transferred to the current National Secretariat for the Environment and Sustainable Development.

The public company regulated the norms specified in the decree and established the permissible limits for dumping wastes at industrial and other premises as well as the limits permitted in absorbing wells and the analytical techniques used by the National Water Works utility to analyse waste dumping at both industrial and special premises (Provision 79179/1990 of National Water and Sewage Company, Official Gazette, August 1, 1990, Annexes A, B, C and D).

The application of the said regulations was transferred to the former Control Bureau for Water Pollution, currently, the Bureau for Pollution Prevention and Management, which reports to the federal Secretariat for the Environment and Sustainable Development (Decree 776/1992, Section 9). This body is in charge of receiving sworn declarations every year from the companies that dump wastes, of authorising the operation of new industries, and collecting the special stamp duty for pollution control (Decree 674/1989, Sections 10, 19 and 6).

Additional regulations were issued setting a total pollution weight limit (value: 1.500) (Resolution 231/1993, Secretariat of Natural Resources and Human Environment), and a total pollution weight limit (value: 80) for dumping wastes from industrial or special premises that contain hazardous substances of an ecotoxic nature (Resolution 242/1993, Secretariat of Natural Resources and Human Environment).

With the aim of avoiding actions that may directly or indirectly cause degradation of the environment and water resources, the Bylaw for Sustainable Management of Sludge Generated in Effluent Treatment Plants was adopted (Resolution 97/2001 of November 22, 2002, Ministry of Social Development and Environment). The recipient body of sludge is the soil, but mud cannot be used for soil improvement in places such as areas covered by snow or frozen, in areas where drinking water is collected, in areas located fewer than 15 meters from the banks of surface watercourses and in hilly areas and areas with extreme rainfall (Ditto, Section 18). All these provisions are limited to the city of Buenos Aires and the districts that surround the metropolitan area and which are part of the licensed services of the former National Water Works utility.

For the Río de la Plata coastal strip adjacent to the city of Buenos Aires and the metropolitan area, the Ministry of Social Development and Environment established environmental quality goals and priority uses, including the supply of conventionally treated drinking water, aquatic life protection, and recreation. Those granted permits to dump wastes must comply with priority quality objectives by 2008 (Resolution 634/1998 of August 16, 1998, former Secretariat of Natural Resources and Sustainable Development).

Since 1992, the water quality programme in the Río de la Plata coastal strip, from San Fernando to Magdalena, Buenos Aires province, has been implemented. Therefore, this programme covers an area from the north of the city of Buenos Aires's metropolitan area to the beginning of Samborombón Bay in Punta Piedras.

Sampling and sediment collections are performed every 500 metres in straight lines that go from the coast for a distance of 10,000m off the coast, and chemical and biological tests are carried out simultaneously.

The Río de la Plata southern coastal strip programme received contributions from the Water Resources Under-secretariat and the metropolitan area's concessionaire, Aguas Argentinas S.A. Moreover, the Naval Hydrographic Service was in charge of fieldwork, taking samples, analysing and assessing results, in collaboration with the Limnology Institute at the University of La Plata.

The guidelines for the assessments on the required parameters, such as BOD, metals and even PCBs in fish, are those adopted by the Administrative Commission of the Uruguay River in 1983 and the Group of Experts in Water Quality from the Inter-governmental Coordinating Committee (CIC) in 1987 (Permanent Committee to Monitor Water Quality on the South Coastal Strip of the Río de la Plata 1997).

Buenos Aires province is on the banks of the Río de la Plata and in its 1999 Water Code (Buenos Aires Province Statute 12.257), it declared that it was entitled to carry out projects in the River with the consent of national authorities (Section 8). Water quality in greater Buenos Aires, especially as regards drinking water, is included in the services provided in the city of Buenos Aires.

Provincial jurisdictions and some municipalities have set environmental regulations that supplement national legislations on the subject. As a result of this multiplicity of jurisdictions, the legal framework is complex and difficult to consult. This inconvenience has been solved by compiling a compendium of the regulations that are spread throughout government, in the Environmental Provincial Digest, issued in 1999-2000 by the Secretary of Natural Resources and Sustainable Development, as part of the Programme of Institutional Environmental Development (PRODIA 2000). The Digest also points out the common basic principles incorporated in the different regulations.

As mentioned earlier, statute 25.688, enacted in 2002, lays out the minimum environmental protection standards for the preservation and rational use of water. Pertinent bodies and authorities, who will be appointed some time in the future, will be in charge of setting parameters and standards for water quality and will "set the maximum pollution limits acceptable for water, depending on its different uses."

Appropriately, the statute includes the regulatory framework of uses and water quality in a statement about the minimum environmental standards. This is necessary for an efficient management of water resources. However, on steps to be taken on water quality the criterion incorporated is not compatible with the principles of environmental policy incorporated in statute 25.675 dated November 6, 2002, especially as regards the principles of prevention and precaution that oppose the concept of the assimilation capacity of the recipient body. In fact, the application of said environmental principles implies an acknowledgement that water is not a recipient of pollutants and that effluents and both domestic and industrial wastes must be sent to treatment plants and sludge must be disposed of according to adopted environmental regulations (Resolution 97/2001 on Sustainable handling of sludge generated in treatment plants for liquid effluents and Statute

25.612, dated July 3, 2002 on Integral management of industrial waste and service activities; foregoing Statute 24.051). Persistent organic, what was set forth in the Stockholm Agreement on Persistent Organic Pollutants, signed in Stockholm on May 22nd, 2001, and inorganic pollutants should not be incorporated and biodegradable substances should receive proper treatment before being disposed of or reused.

13.3.2 Water Quality for Human Consumption

In Argentina, water quality regulations are laid out in the Argentine Food Code that explains which physical, chemical and microbiological characteristics have to be satisfied. Parameters and guidelines are adopted as a result of mainly following the standards set by the joint Food and Agricultural Organisation (FAO), World Health Organisation (WHO), Food Standards Programme Codex Alimentarius Commission and other international sources (Section 982, Drinking water, Argentine Food Code, Chapter XII, Statute 18.284 dated June 18, 1969, updated by Resolution 474/1994 of the Ministry of Public Health). Nevertheless, it is necessary to adapt these quality standards to local conditions and to the treatment technologies applied locally. The Code authorises the Health Authority to have jurisdiction to admit different values from those set, if necessary, due to the normal composition of the water and the impossibility of applying corrective technologies.

So that national authorities could count on a methodology to formulate drinking water quality regulations that guaranteed inhabitants that their water was safe and would be in the future, the Under-secretariat of Water Resources and the former National Institute for Water Science and Technology, currently National Water Institute, prepared a "methodology to establish local regulations on water quality for human consumption" (Goransky and Natale 1996). This document is a non-applicable regulation but it sets out the basic criteria for the State and different provincial jurisdictions to establish regulations on water quality in an unified way so as to protect public health while taking technological feasibility into account.

The report includes guideline based on the use of data applied by WHO and the Environmental Protection Agency of the United States of America (EPA), complemented by local information and the special characteristics of the population that uses drinking water. To produce the report, information was collected from different sources and stored in three databases. The first is a database of guidelines and quality standards for water for human consumption, with data coming from 13 national and international reference sources. The second is a toxicological database, which includes basic toxicological information on 507 chemical parameters (System IRIS, U.S. EPA 1993 and WHO). The third is a database on the quality of supply sources, which includes information on water quality from the sources, supply water quality and the technological characteristics of waterworks operating in the provinces of La Pampa and Santa Fe, as well as in the metropolitan area of Buenos Aires. The report also points out the need to count, among other guidelines, on information about the characteristics of the sources of water supply to be able to use the methodology created. Regarding supply sources, guidelines are available for both surface and groundwater.

The conclusion of the methodological bases on drinking water guidelines is a decisive step towards introducing national regulations on water quality. An analysis of the risks associated with chemical parameters and protecting the health of the population is also included.

Adulteration of drinking water is an attack on the health of the population and it is a crime, according to the Argentine Criminal Code (Statute 11.729 from 1921, as amended). Section 200 of the Code states that "anyone who poisons or adulterates drinking water in a way that threatens human health shall be imprisoned for between three and 10 years ... If a person dies as a result, the sentence shall be between 10 and 25 years' imprisonment." Pollution was incorporated in the legal definition of a crime in Decree-statute 17.567/1967 that increased penalties for whoever "poisoned or adulterated" water adding those who "poisoned, polluted or adulterated drinking water in such a way that it threatened health." The amendment was superseded upon enactment of Statute 20.509 in 1973 that returned to the original wording of the provision; therefore, currently the act of "polluting" without further consequences is not considered a criminal offence.

The legal definition of a crime is that not only the act performed with intent or malice is punishable but also the act "performed by imprudence or negligence or by lack of expertise or by non-compliance with regulations or ordinances." As regards these suppositions, there are pecuniary penalties if, as a consequence of the punishable behaviour, the act does not cause sickness or death among humans. On the other hand, if an unintentional crime causes sickness or the death of the people, punishment that deprives the guilty party of his/her freedom will be applied (Section 203, Criminal Code).

When considering the measures that must be adopted to protect drinking water supplies, the treatment of home and industrial effluents in both surface and groundwater must be pointed out. Drinking water supplies for the general population in Argentina are deficient, although not as deficient as in the other South American countries. One of the difficulties in increasing the number of dwellings having access to running water and sanitation services is the high cost of installing different technologies and the necessity of having specialised and permanent maintenance, as well as raw materials that in many cases are not present in the local market (Ingalinella et al. 1997; Di Bernardo 1993).

Relying on low-cost, user-friendly technologies is an option that may help expand, in an exponential way, the protection of water quality through the proper treatment of liquid wastes. In Argentina, following the Brazilian example, prototype treatment systems have been developed. They use efficient technologies with low installation costs, easy to operate and maintain and especially suitable for medium-size and small towns.

Priority has been given to the pre-treatment stage that, where possible, does not use chemical products, and to the slow filter systems, also used in the United States and Europe, that facilitate the final stage by eliminating solids beforehand. The methods of horizontal gross filtration and ascendant filtration in gravel layers have been successfully applied in medium and small towns. In densely populated cities such as Rosario studies have been carried out to modify the regulations on water quality in the province of Santa Fe because of a lack of living beings, phyto and zooplankton.

In the Buenos Aires metropolitan area, before the drinking water and sanitation concession was awarded, the technical requirements to meet the required service levels were laid out (Decree 999/1992, dated June 18, 1992, Section 42). Minimum regulations were introduced to cover the quality of the water produced the year the concession started and for the following 10 years. The system and the frequency of sample collection by the concessionaire were regulated. The concessionaire is in charge of "establishing, operating and registering a sampling regime, both for a normal service and for emergencies. The sampling is of raw water, water undergoing treatment and treated water, with the purpose of controlling the water all along the supply system"(Section 42b). The parameters were established for quality control, both of raw water from surface and groundwater inlets and treated water coming out of the waterworks and the distribution system. Furthermore, regulations were issued for sewage, drainage and the control of liquid wastes dumped by the service concessionaire into the water courses, where the parameters for sewage and industrial drainage were determined, depending on the established frequency. The indicated frequency is three times a year in drainage to drainpipe and twice a year in drainage to recipient bodies. In Argentina about 70 percent of the urban population receives drinking water and sanitation services through private or mixed companies that are awarded concessions.

The Water Regulatory Body is in charge of regulating and controlling the quality of the water and sanitation services rendered by the metropolitan area concessionaire and other service providers (Statute 23.696 and Decree 999/1992, dated June 18th, 1992, Annex I, Sections 13-14).

The city of Buenos Aires as well as 17 municipalities of the Buenos Aires province which constitute the metropolitan area of the city of Buenos Aires, have licensed the water and sanitation services, along with other cities of the province of Buenos Aires. The licensees are regulated according to the quality of the service provided (Statute 11.820, 1996, regulatory framework for water and sanitation services in the Province of Buenos Aires). A regulatory body has been created, the Regulatory Agency of Buenos Aires Waters, an autonomous entity responsible for regulating, controlling and supervising utilities (Decrees 743/1999 and 2307/1999 of the Province of Buenos Aires) under the conditions established by the Water Code of the Province of Buenos Aires (Statute 12.257, 1999). To provide the services for many cities in the province that do not have licensed utilities, the mainly state-owned company Buenos Aires Waters has been created.

13.3.3 Ambient Water Quality

The development of water quality guidelines is one of the substantive functions of the Under-secretariat of Water Resources. In 1997, it called for two joint working commissions, made up of representatives of the Under-secretariat, National Water Institute and Secretary of Natural Resources and Sustainable Development, to work on the subjects of water basins and water quality. The coordinating commission for matters of water quality suggested that a start be made on establishing national criteria for ambient water quality in the same way as quality criteria for treated water were set. Once the project was defined, the ad hoc working group, which started its task in March 1998, drew on different professional disciplines. The programme concept is to protect aquatic life. Since 2003, the programme, which has undergone internal and external audits, has become permanent and it reports on its activities twice a year (http://www.obraspublicas.gov.ar/hidricos/calidad_del_agua_actividades.htm).

National guidelines for ambient water quality are set after consideration of the following: sources of the surface and groundwater supply system for human consumption, crop irrigation, recreation and livestock, as well as the protection of aquatic ecosystem.

The programme is a management instrument aimed at designing the minimum standards of environmental protection. The principles of environmental protection must have a framework and, as a consequence, the criteria are calculated for the first time by employing a methodology for each type of water use. It is a permanent task because guidelines are based on scientific information that is modified and therefore the guidelines are changed. The guidelines are for protecting water resources that makes it possible to establish regulations for discharging wastes and emissions that adhere to the guidelines as closely as possible. Quality goals have to be associated with schedules that are demanding but achievable.

It is expected that the concept will be included in a future water management statute and that the proposed values will be adopted as national guidelines, replacing those adopted in 1987 for the Río de la Plata Basin and those included in Statute 24.051. Mechanisms for implementation must also be established because if there are neither measurements nor assessments there will be no control of pollution levels or water quality. This does not imply overlapping jurisdictions or bypassing provincial powers. Guidelines are not regulations but they do contribute substantially to the whole process.

The provinces have not yet established water quality criteria. Tierra del Fuego province, which has not produced any criteria either, has regulations limited to coastal waters. More modest goals were adopted for the Yacyretá dam and reservoir in the Paraná River. The goals are similar to those adopted by the Brazilian agency CONAMA. They are less restrictive than those of the 1987 Document for the Río de la Plata Basin.

The National Water Institute, an agency of the Under-secretary of Water Resources, is a decentralised institution of science and technology that carries out many tasks linked to water resources through its specialized centres. It was set up in 1973 as National Institute for Water, Science and Technology, devoted to the study, research, technological development and consulting services regarding development, control and conservation of water. It covers hydraulics, hydrology, sediments, water quality, treatments and others (www.ina.gov.ar).

One of its centres is the National Programme for Water Quality, devoted to developing and applying methodologies to monitor and simulate the transportation and destination of pollutants in water systems, as well as assessing and managing the risk they imply; determine scientific and technological criteria, guidelines and regulations on ambient water quality, water for priority uses, sediments and water biota. One of its projects is the Assessment of Risk Factors caused by pesticides in the rural environment, and one of its permanent activities is to work on setting national guideline levels for ambient water quality.

Another of the specialised areas is the Technological Centre for the Use of Water, which conducts research, development and technical assistance projects that are related to water quality in recipient bodies, technology for water treatment and treatment and disposal of wastewaters and sludge, water quality of recipient bodies; analytical chemistry of water pollutants, sediments, water biota and sludge of different origins and water toxicology. It also tests different electromechanical equipment for water treatment and the operation of treatment plants at different levels, among other activities.

The important Centre for Water Economy, Legislation and Administration that operates from the city of Mendoza, is part of the National Water Institute. Its purpose is the institutional progress in the sustainable management of water resources. It contributes to the design of economic, administrative and legal instruments for balanced water management, according to the guidelines for administrative efficiency, recovery of investment, operation and maintenance costs as well as users participation. This Centre also performs advanced training tasks, using an interdisciplinary approach aimed at political, managerial and operative levels of those institutions concerned with management of water resources.

The National Water Institute is not empowered to suggest or apply regulations, but it acts in its organisational capacity, concentrating on studies that lead to technical determinations, assessments and information processing.

13.4 Some Case Studies

13.4.1 Basin of the Matanza-Riachuelo River

This basin, now completely urbanised, has a surface area of 2,240 km² and is on the border between the city of Buenos Aires and Buenos Aires province. The Riachuelo, a historic location, was the original port of the city and the destination for ships that crossed the Atlantic Ocean to reach this calm tributary of the Río de la Plata. It is a water basin that gently slopes down into formerly fertile prairies. It has a mild climate, plentiful rains and good wind patterns, important urban centres and infrastructures, and a productive tradition that ranges from agriculture and livestock to industrial activities, as well as an important petrochemical centre in the Dock Sud (South Dock). Although this scenario offers attractive opportunities for sustainable development, its thoroughly polluted waters and environmental degradation adversely affect the quality of life of its inhabitants.

The basin comprises a main course called Matanza in Buenos Aires province and Riachuelo when it flows through the outskirts of the city of Buenos Aires and reaches the Río de la Plata, and several low-flow courses that drain towards its stream. More than three million people, eight percent of the country's population, live in the basin. Water in some stretches has only 0.5 mg/BOD/l, well below the 5 mg/BOD/l necessary for animals and plants. This is why it is unsuitable for treatment or performing most of the activities generated by the presence of a watercourse. The degradation of water quality started in the 19th century, when the export of meat, leather and wool was conducted from its banks and slaughterhouses sent effluents straight into the basin. Late that century the pollution was already significant and it started an epidemic that caused thousands of deaths in the city. This tragedy brought little change because, a little while afterwards, the meat-processing industry began to inhabit its banks and the situation grew even more serious. The industrialisation process that took place during the second half of the 20th century and which spread across the Buenos Aires metropolitan area, increased the amount of pollutants because untreated industrial and domestic effluents were thrown into its waters.

Despite being short, the basin comes under the jurisdiction of the Nation, the city of Buenos Aires, Buenos Aires province and several municipalities. Therefore, it is inevitable that every initiative to carry out an environmentally sustainable management plan for the basin is structured on inter-jurisdictional grounds. In 1995, Decree 482/95 created the Executive Committee for the Environmental Management Plan and the Matanza-Riachuelo River Basin Management on which the national government, the city of Buenos Aires and Buenos Aires province are represented. The task of this Executive Committee is to launch the Environmental Management Plan and the Matanza-Riachuelo River Basin Management to be designed as a remediation process by the Executive Committee for the Cleaning up of the Basin (Decree 1093/1993). The Executive Committee reports to the National Under-secretary of Water Resources (Decree 27/2003).

The Plan received financial support from the Inter-American Development Bank (Environmental Management of the Water Basin of Matanza-Riachuelo - AR-0136, Decree 145/1998, granting USD 250 million during the implementation of the programme and an identical counterpart from the Argentine government) and the donation of a laboratory for chemical analysis through a cooperation project between the Japanese International Cooperation Agency and the National Water Institute. The Plan was complemented with technical cooperation providing experts from Japan to collaborate with local institutions.

The studies carried out have made it possible to diagnose the situation of the basin and the programme of actions to be carried out regarding water quality, toxic risks, measures to reduce the harmful effects of floods and threats to the population's health. The flow is excessively slow, mainly due to sea vessels and wastes thrown onto its bed. The task of removing abandoned hulls has made some progress, although there is still a lot to be done (Decree number 180/99).

The Special report on the Matanza-Riachuelo River Basin (Ombudsman's Office et al. 2002) coordinated by the national government, analysed thoroughly the background of the basin, its then situation and the results of the programmes in progress. Upon looking for the causes of the obvious lack of progress what stands out most is that the body responsible for the performance of the cleanup programme lacks the institutional strength to take the necessary measures. As a matter of fact, it does not have powers to control the activities carried out in the basin. It depends on the measures taken and performed by provincial and municipal authorities. Some 3,000 industrial premises have been surveyed in the area, including 100 of the biggest polluters. The most serious problem is the danger to the health of the population, especially that of children, because of the poor living conditions and toxic risks. A greater commitment from the responsible authority is necessary to correct the high urban impacts and to carry out the works necessary to restore a healthy environment.

13.4.2 Colorado River Basin

The Colorado is a patagonian river, and belongs to the Atlantic watershed. It is formed by the confluence of two rivers that have their source in the Andes, the Grande and Barrancas Rivers. It crosses the whole patagonian plateau in its course until it reaches the Atlantic Ocean. It is 1,200 km long and goes through the provinces of Mendoza, Neuquén, La Pampa, Río Negro and Buenos Aires. The Casa de Piedra dam has been built on this River. The drainage basin covers an area of 15,300 km² and the yearly average flow, 25 km away from its source at Buta Ranquil station, is 148,6 m³/s (period 1940 - 2002).

Since 1956, meetings have been held with representatives of the five riparian provinces for the development of its water resources, which is in shorter supply than before due to low rainfall levels. In 1976, the Inter-jurisdictional Committee for the Colorado River was created. This Committee was to be in charge of the Programme to Enable Areas for Irrigation and Distribution of the Colorado River Flow. Oil exploration activities, irrigation and drinking water supplies for urban centres are developed in the basin.

Since 1977, an Integral Programme for Water Quality (Quality of the Aquatic Environment Sub-programme) has been in operation, making it possible to know the status of its water quality, bed sediments and fish. Research is carried out according to internationally approved protocols. Monitoring of metals (arsenic, cadmium, zinc, copper, chromium, mercury, molybdenum, nickel, lead and selenium) and polyaromatic hydrocarbons (HAPs) and chronic eco-toxicological tests are performed. Qualitative and quantitative determinations of aliphatic hydrocarbons have also been performed. In the sediments, especially in the Casa de Piedra reservoir, research into the presence of metals and hydrocarbons is conducted. The presence of toxic substances in fish is being analysed to ascertain if it poses a risk for human consumption.

Water and external temperature, electrical conductivity and pH are determined in the eight river gauging stations. The results obtained are compared with guidelines (WHO 1993, CCME 2002, Canadian Water Quality Guidelines for the Protection of Agricultural Uses, Irrigation, Livestock, and for the Protection of

Aquatic Life) to assess water quality before considering it as a source for drinking water, irrigation and livestock, as well as for the protection of aquatic life.

The results indicate that in the areas of study the Colorado River (Integral Programme on Water Quality of the Colorado River, Quality of the Water Environment, 2002:42) water is still suitable for drinking. Very little HAPs presence was detected, and when such a presence was detected, it was always below the standards set for the programme, which confirms the resource is still suitable for maintenance of aquatic life.

As regards sediments, the presence of metals and metalloids was observed in sediments at the bottom of the Casa de Piedra reservoir and downstream in the Colorado River. Concentrations of some parameters in the reservoir increased between 2000 and 2002, although this increase is considered to be related to the geology of the area and not due to human activities. The concentrations of arsenic, cadmium and copper in the reservoir sediments slightly exceed the guidelines but they remain below those of the river (CCME 2002). There are no HAPs in the sediments at the bottom of the reservoir and in the few cases that they were detected, their levels were lower than the guidelines established for the protection of aquatic life.

As for the fish, results show that it is unnecessary to recommend restricting its consumption by humans, according to applicable regulations (Resolution by the National Service of Health and Food and Agriculture Quality in 1994; EPA 1999) because it entails no risk to human health.

The creation and performance of the Water Quality Programme for the Colorado River shows that depending on a network that works properly makes it possible to observe water quality and take corrective actions whenever necessary. The efficiency of the regulations issued has been confirmed by the fact that the application of the programme has started a mechanism that makes it possible to optimise uses, oil installations, agriculture and consumption, and to control harmful effects.

13.4.3 Salí-Dulce River Basin

Located in the north east of Argentina, between parallels 22° and 28° S, the Salí River Basin is high seasonal by nature and is the main source of water for the endorheic basin, Laguna de Mar Chiquita.

The 42 affluents of the Salí River cover almost the whole of the rural area with the highest population density in Argentina, Tucumán province (50 inhabitants per km²) and a small part of the provinces of Catamarca and Salta. Its waters descend from a height of 5,500 metres from the Nevado of the Aconquija mountain, with 1,800 mm of yearly rain, to the Santiago del Estero plain, which it crosses diagonally, before flowing into the Laguna de Mar Chiquita, at 62 m above sea level, in the province of Córdoba. The watercourses feed two big reservoirs, one at El Cadillal and the other at Hondo River, that cover an area of 33,000 ha, after which they run through the bottom of the basin as the Dulce River. Many projects for irrigation, water supply, industrial activities, and so on, have been built in the basin.

The Hondo River reservoir (1967) made it possible to enlarge the irrigated area. This gave rise to significant expansion of the agricultural area, by about 200,000 ha.

On the Santiago del Estero plain, floods cause marshlands when the river level rises, which enable seasonal agriculture that has been carried out since the pre-Columbian times. Before reaching the Laguna de Mar Chiquita (Mar de Ansenusa) the river strays into marshlands, saltpeter mines and ponds, all of which were designated as a Ramsar wetland protected area in 2002.

The plain of the province of Tucumán has undergone intensive sugarcane development, with smallholdings, an industrial sugar refinery and citrus orchards, a paper industry and both permanent and seasonal towns that have urbanised the area. In the last two decades, a change in land use has taken place, with increased citrus plantations and other fruit and vegetable farming. Nevertheless, the pollution situation has not improved and untreated effluents have caused serious eutrophication in the Hondo River reservoir. This is detrimental to the downstream uses of the reservoir in Santiago del Estero province, which has made claims over the issue and has started legal action. Due to the erosion caused by the inappropriate land use, the Salí River has departed from its natural course and a delta has formed at its mouth. Furthermore, the silting up level of the reservoir has reached a sedimentation level of 2 percent annually in recent years.

Projects regulating the dumping and treatment of effluents, land use, sanitation, water quality and sedimentation controls have not solved the problem. There is no basin authority and the provincial authorities have not studies the situation carefully.

13.4.4 Neuquén, Limay and Negro River Basin

The Negro River, located in Patagonia, is formed by the convergence of the Neuquén and Limay Rivers. It is a basin of 116,000 m^2 that runs through the provinces of Neuquén, Río Negro and Buenos Aires in its final stretch, before flowing into the Atlantic Ocean. There are important hydroelectric operations with an installed capacity of 4,200 MW and a yearly average power generation of 13,500 GWh. The hydroelectric potential is estimated to be 33,000 GWh per year (Instituto Argentino de la Energía 1999). The sub-basin of the Limay and Neuquén Rivers is especially rich in hydrocarbons that are transported by pipelines to refineries in La Plata and Bahía Blanca and by gas pipes to cities in Argentina and Chile.

The Negro River Valley has an abundance of fruit and produces about 1,200,000 tonnes annually in an irrigated area of 150,000 ha. Livestock is also bred there.

Since 1970, water quality has deteriorated due to an increase in population and the resulting discharge of untreated household wastes into the water. Another problem is the discharge of matter caused by erosion in mountain areas a result of deforestation. Work to control river levels has altered the unregulated regime and self-purification capacity of the basin, which in turn has affected the ecosystems.

Hydrocarbon exploitation has provoked extensive soil pollution and created open-air sinks. It has also had impacts on both surface and groundwater.

In 1985, the riparian provinces and the national government agreed to create a basin authority, the Inter-jurisdictional Authority for the Basins of the Limay, Neuquén and Negro Rivers (AIC), established in 1991. Regarding water quality, the functions of AIC are to adopt regulations and take action to prevent, avoid and correct polluting processes, carry out research, implement projects and build, install, operate and give maintenance to installations in order to detect and control water pollution in the basin.

Since 1993, the organisation has carried out programmes to improve the environmental aspects of the basin in terms of water quality, covering environmental assessment, supervision and planning of the water resource and other water-related resources. The programmes include survey of wildlife, monitoring and control of fish, the survey of birds, studies of the impact of human activity on water bird communities, studies of the impact of oil activities on areas next to the river, monitoring and control of water quality in rivers and reservoirs, qualitative diagnosis and zoning of pollution in the rivers of the basin, the environmental management plan for lakes in the basin, the study of algal bloom in reservoirs, and pollution assessment for aquiculture in reservoirs.

Operating these programmes has made it possible to follow up on water quality conditions and to prevent the negative impacts of certain uses. Monitoring is carried out with the support of a network of gauging stations, making possible to supervise the parameters for water quality. AIC counts on funds from private hydroelectric concessionaires and consequently can perform its duties in autonomously (Bylaw of the Inter-jurisdictional Authority of the Basins of Limay, Neuquén and Negro Rivers, Section 19c). It is a positive example of institutional progress in the sense that it is possible to control pollution, and apply regulations for water quality protection.

13.4.5 Outcome of Case Studies

The cases studies of the four basins in Argentina described above represent different attitudes towards the challenges inherent in water quality management. While regulations have been adopted through a network that operates efficiently in the case of two big basins, those of the Colorado and Negro Rivers in Patagonia, in smaller but more densely populated areas such as the Matanza-Riachuelo River, the metropolitan area of Buenos Aires, the Salí-Dulce River area in the Provinces of Tucumán and Santiago del Estero, no measures have been taken to reverse the high levels of pollution of the water bodies.

By comparing these examples one can infer that there are legal and institutional mechanisms available to carry out sustainable water management. The benefits of preserving and cleaning up the resource can be objectively assessed. In fact, it is possible to prove just how efficient regulations are.

The decision to create tools and use them is something laws and institutions cannot influence. Undoubtedly, more vociferous complaints from people may bring about a change of attitude in the authorities in those areas that are environmentally and socially neglected.

13.5 Water Quality Issues in Río de la Plata Basin

Watercourses in Argentina are mainly international. In the Río de la Plata River Basin, shared with Bolivia, Brazil, Paraguay and Uruguay, 84 percent of the country's water resources are concentrated. The riparian countries, through the Río de la Plata River Basin Treaty, Brasilia 1969, created an institutional system whose purpose was to use the water resources by developing the infrastructure of the basin. The treaty established the Meetings of the Ministers of Foreign Affairs of the Countries of the del Plata Basin, which is the decision-making body. Meetings are usually held at least once a year or whenever appropriate. The CIC, made up of political and technical representatives of the riparian countries, complies with the decisions approved by the Meeting of Ministers of Foreign Affairs, meetings which are held on a regular basis. The General Secretariat operates permanently and performs the activities requested by the CIC.

Regarding the protection of water resources, the first objective of the riparian countries was to be informed of water quality conditions in the main watercourses, among which are the following rivers: Paraná, Paraguay, Uruguay, Pilcomayo, Bermejo and Iguazú. The importance of this subject is crucial for the integrated management of the basin, which, because of demographic growth and agricultural and industrial developments, is undergoing the related phenomena of disorganised urbanisation, deforestation and loss of plant cover that accelerate sedimentation, pollution and degradation of water resources. There are also other cases that deserve special attention, such as mining, that cause serious pollution in some terms of certain effluents, for example, in the Pilcomayo River Basin.

As a result of the formation of the Meetings of Ministers of Foreign Affairs, the CIC established a system of regular meetings for the five member countries. Since the 1970s, the experts of each country have met regularly. They have agreed to exchange information on established parameters. Monitoring stations installed in each of the countries were integrated to operate as a network. Even though the dimension of the network could be considered insufficient in relation to the extension of the drainage basin, the information taken into consideration came from 28 stations in a territory of 3 million m^2, it was an important starting point for setting up a system of permanent communication. It has yielded encouraging results. Currently, only 18 stations are in operation but measurement of the parameters has not been reduced because each country has improved technology at a national level.

In 1976 a decision was taken to start testing in order to determine the real status of the quality of water. The purpose was to launch joint actions which might help resolve or minimise pollution problems. During the first semester of 1977, governments were asked to send proposals for listing minimum parameters, that should be investigated and the methodology for establishing them (Resolution 67, VIII) dated on December 9, 1976). As a result, a group of experts in water quality

(Pollution Subgroup) was formed. After fruitful exchanges and interactions, it submitted its report on October 19, 1977. The conclusions of the technical experts were approved by Resolution 90 (IX) dated December 9, 1977, of the Meetings of Ministers of Foreign Affairs of the Countries of the del Plata Basin. The elements to be determined were those of "temperature, turbidity, PH, conductivity, chlorides, dissolved oxygen, arsenic, cadmium, mercury, cyanide, copper, chromium, lead, zinc, coliform bacteria (total and faecal) and BOD (Biochemical Oxygen Demand)" as well as "other parameters that local conditions may indicate and demonstrate harmful changes from the health, ecological and/or economic points of view of the water resources." It was agreed to start a programme for assessing water quality in the rivers of the basin on January 1, 1978, with a minimum network of gauging stations.

The request for exchange of information has been repeated in Resolution 123 (X) dated December 6, 1978, and related Resolutions 192 (XV), 196 (XVI) and 2 (E-II). In Resolution 140 (XI), dated December 4, 1980, the CIC was asked to call the Subgroup Water Qualities for the purpose of: a) determining the parameters of water quality that should be incorporated; b) adopting the suitable analysis or determining methodology; and c) selecting one or more reference laboratories in each country.

As a consequence, experts submitted different reports based on this process. These were presented at meetings of Technical Counterparts, held for the first time in October 1986, and then in April 1987. During this second meeting, the Argentine delegation submitted the work document that established water quality guidelines for the first and last time. The guidelines in that document are still used as a reference, although they have not been adopted as a regulatory instrument at a regional level (Argentine Delegation, 1999, Work document on Water quality Assessment and Pollution Control, 1987-1998 submitted at the Second Meeting of Technical Counterparts of the del Plata basin in April 20-24, 1988). Therefore, it should be noted that the water quality activity regionally has been important for Argentina, even though it has been very limited.

The Minimum Network of Gauging Stations worked continuously in the Argentine sector of the basin from 1987 to 1995, and intermittently and with fewer stations until 1998. In the beginning, the network coordinated by the Undersecretariat of Water Resources had 13 stations, run by different national and binational bodies. Three others were added later. However, due to the different origins of the bodies in charge of them, not all stations operated during the whole period, and not all the selected parameters were determined in all of them. As a matter of fact, the network incorporated stations that belonged to binational, national and provincial bodies that are still operating, even though they are not part of a network.

The Yacyretá Binational Entity runs its own monitoring network with about 20 stations from Itá Ibaté (waters below the reservoir) to Puerto Libertad (almost at the mouth of the Iguazú River). The Administrative Commission of the Uruguay River operates its own network as does the Salto Grande Joint Technical Commission for the dam.

The Administrative Commission of the Río de la Plata participates with the Joint Technical Commission of the Maritime Boundary of the FREPLATA Project which, since its establishment in 1999, has generated, received and processed information about water quality in the Río de la Plata (FREPLATA Project, no date).

In Argentina the Naval Hydrographic Service is the body that reports on water quality in the Río de la Plata and is in charge of the sampling and monitoring of the river's coastal area from San Fernando to Punta Lara (South Coastal Project).

The Tri-National Commission of the Pilcomayo River, Argentina, Bolivia and Paraguay, has a gauging station, located at the same latitude as the village of Misión La Paz, in Salta province. Information on water quality, with samples of relative frequency obtained from a limited network, is also obtained in Tarija. Nevertheless, the main role of the Pilcomayo River was developing water resource usage, including water quality aspects. From 1998 to 2000, a programme was carried out to determine the toxicity of metals in living resources in the courses of water where the National Institute for Water, National Commission of Atomic Energy and Miguel Lillo Foundation from the University of Tucumán participated. In the sources of the Pilcomayo River, there is a historic pollution process caused by mine tailings that can be partially solved by building waste reservoirs specifically for mining operations. Such reservoirs are about to be built, although other problems originate from overflows from existing reservoirs because of heavy rains.

In December 1987, the XVII Meeting of Ministers of Foreign Affairs adopted a programme of actions (PAC) which included the recommendations of the II Meeting on Technical Counterparts on Water Qualities (Project II − 1, Resolution 203 (XVII) adopted at the XVII Meeting of Ministers of Foreign Affairs, 1987). In their Report, the experts agreed on the importance of increasing technical and operative cooperation and of counting on an installation network in order to carry out continuous assessment of the recommended quality parameters. In compliance with this, the installation and start-up of a minimum network of gauging stations was declared a priority. This network would make it possible to perform systematic and periodic controls in order to determine the pollution of the water in the international rivers of the basin and to have a global situation outlook (Resolution 206 (XVII) adopted at the XVII Meeting of Ministers of Foreign Affairs, 1987).

In the third meeting of experts, held on July 1 1988, progress was made on such issues as new parameters to assess. The Meeting of Ministers of Foreign Affairs received the conclusions of the report on "Parameters to Assess" and decided to recommend that governments adopt the listing of the new parameters to be included for the assessment and exchange of information, viz: ammonia nitrogen, nitrate nitrogen, nitrite nitrogen, phenolic substances, total iron, organochlorine and organophosphorated pesticides, anionic detergents, chemical oxygen demand, hydrocarbons, total phosphorus and suspended solids (Resolution number 221 [XVIII] from 1989).

In 1991, the Methodological Guide for the Operation and Assessment of the Network of Water Quality in the del Plata Basin (Annex to Resolution number 233 (XIX) 1991) was approved so that there could be a common element for reference when making the "diagnosis of the status and assessment of the quality levels of the waters that ensure the representative nature and reliability of the values obtained." The objective was also to contribute to the World Conference on Environment and Development, in Rio de Janeiro, in 1992 (Resolution 233 (XIX) adopted at the XIX Meeting of Ministers of Foreign Affairs, 1991).

The report in that Meeting of Ministers of Foreign Affairs to assess Project II - 1 (PAC) confirmed the partial operation of the Minimum Network of Gauging Stations for the assessment of water quality, which counted on 15 of the 27 scheduled stations, most of them in Argentina and shared by Argentina and Paraguay on the Paraná River and by Argentina and Uruguay on the Uruguay River (Resolution 232 (XIX), adopted at the XIX Meeting of Ministers of Foreign Affairs, 1991).

Furthermore, due to important pollution issues in the Paraguay River and some of its tributaries, in 1991 a resolution was adopted enabling CIC to give a boost to joint studies on water quality in the sub-basins for the purpose of making "an environmental diagnosis that includes the identification and description of pollution sources and an assessment of the water quality background corresponding to the following parameters: dissolved oxygen, biochemical oxygen demand, chemical oxygen demand, total phenol, nitrogen series and suspended solids in the aquatic environment, in suspended solids, in sediments and in fish tissue" (Resolution 235 (XIX), adopted at the XIX Meeting of Ministers of Foreign Affairs, 1991, paragraph 1.a).

The proposed programme ought to have complemented the Study on Environmental Impact that the project for the Paraguay-Paraná Waterway (Cáceres, Mato Grosso, Brazil, to Nueva Palmira, Uruguay), which was incorporated at that same meeting as an organ of the institutional system of the del Plata Basin, was to carry out (Resolution 238 (XIX), adopted at the XIX Meeting of Ministers of Foreign Affairs, 1991). To achieve this, support of the regional financing body (Regional Financing Organisation of the Basin Countries, FONPLATA) was to be requested, due to the fact that "granting the basin the necessary means to improve the control and preservation of water quality" was considered a priority (Resolution 235 (XIX), adopted at the XIX Meeting of Ministers of Foreign Affairs, 1991, paragraph 2). Furthermore, it was decided that groups of experts would exchange information "on controlling the use of mercurial compounds and toxic products used in agriculture in each country," taking into account the interest of the member countries in the preservation of water-related wildlife (Resolution 236 (XIX), adopted at the XIX Meeting of Ministers of Foreign Affairs, 1991).

Finally, it was resolved that the same technical counterparts should establish a system to advise on pollution incidents caused by the spillage of chemical substances within the scope of the del Plata basin, in order to diminish the harmful effects of such incidents on aquatic life and the environment in general (Resolution 237 (XIX), adopted at the XIX Meeting of Ministers of Foreign Affairs, 1991).

The following year there was a rise in cholera cases in Argentina and Bolivia and this caused preventive measures be taken in the system of the del Plata basin to avoid the spread of *Vibrio cholerae* in the water biota. An exchange of information for its detection in domestic networks was carried out. The data was included in a document that also indicated the infrastructure of basic cleaning-up in each of the countries on the basin, in addition to coverage indicators, operational conditions and financing needs (Resolution 247 (XX), adopted at the XX Meeting of Ministers of Foreign Affairs, 1992).

During a meeting that took place in May 1999 on water quality and hydrological warnings, the Argentine Delegation submitted the Final Report on Water Quality 1999, regarding 12 stations of the Río de la Plata basin, not only within Argentine but also with shared information from Paraguay and Uruguay. The report includes the data for 1987-1998, and it is considered a valuable contribution for the period it covers and for the amount of information it includes (Argentine Delegation, 1999, Work Document on Water Quality Assessment and Pollution Control, 1987-1998. Document No. 021, submitted at the Second Meeting of Technical Counterparts of the del Plata Basin held April 20-24).

The last meeting of the group of experts took place in Foz de Iguazú in November 1999.

In 2002, in the Identification of Objectives of the System of the del Plata Basin (Decision number 1/02, June 27 2002), the CIC again pointed out that water quality was of permanent interest for member countries. It proposed integrating the activities of member countries into a coordinated monitoring system of the basin watercourses, reconstructing the hydrological network. This objective was incorporated into the Action Programme of the Río de la Plata Basin, approved in November 2002 at the Third Meeting of the CIC Project Unit on November 21, 2002. The infrastructure installed in each of the riparian states "shall determine the physical-chemical and bacteriological quality variables of the water, and this technical information shall be transmitted to the processing centre in charge of taking the results obtained and making them available to those interested. The programme shall start by surveying and assessing the location and features of gauging stations, as well as the laboratories available to determine the set parameters."

At present, water quality objectives of the basin and the mechanisms for their assessment constitute one of the activities of the project, defined as a "Framework for the Sustainable Management of the Water Resources of the la Plata Basin, with respect to the Hydrological Effects of Climatic Variability and Change," an ambitious programme that is coordinated by the CIC with the support of the Global Environment Facility, at the request of the United Nations Environment Programme, with the Organisation of American States as the executing agency. The FONPLATA and other sources fund the project, which started in November 2003.

The regulations adopted by regional and binational commissions with jurisdiction over international watercourse, made up of representatives of competent government agencies, become domestic law when approved at national levels. In turn, government bodies with proper jurisdiction also conduct the bilateral and multilateral joint programmes.

13.6 Conclusions

The focus on the management of water is mainly placed on its uses, to the detriment of the protection and quality of the resource. This can be verified in the new provincial water codes, such as the one enacted for Buenos Aires province in 1999, which, despite having the purpose of protecting the water resources of the province, does not contain a special chapter on water quality and its protection.

Preserving water quality, however, is essential for the health of the population. Hence, the economic cost of taking measures to protect water quality must be properly evaluated along with the cost of treating diseases that originate from water. A proper policy on water quality protection directly effects health, improves quality of life and lower medical and hospital costs.

Behind the aim of adopting water quality standards is the maintenance of the capacity of the water to sustain the aquatic life. One of the reasons why it is difficult to switch attention of water management to quality standards is because there has been a failure to make people aware of the goals being sought.

The aim of establishing and maintaining water quality for specific uses and regions should be clearly established. This is an extremely complex task as the uses vary from region to region. Resource availability, water flora and fauna and sedimentation also have to be considered, among other considerations. There is a need also to make clear that the parameters adopted respond not to current characteristics but to a desirable and achievable model and that they are suggested in order for goals to be accomplished now and in the future.

The economics of it all must also be determined because no decision is taken unless it has a beneficial effect for the agency that decides to implement a policy of protecting water quality standards. Different organisations at national and international levels establish guidelines for water quality that can be, and have been, adopted as general recommendations for water quality management in the region.

In conclusion, protection of water quality should be an important water management goal; sustainability of the biota should be a reason for having good water quality standards; and identification of the best organisations and agencies is necessary so that they can be used as models for proper water quality management.

13.7 References

Di Bernardo L (1993) Métodos e Técnicas de tratamiento de agua, Vol. I y II. Associaçao Brasileira de Engenharia Sanitaria e Ambiental, Rio de Janeiro

EPA (1999) Environmental Protection Agency. http://www.epa.gov/water/

FREPLATA Project, PNUD/GEF/RLA99/G31. Environmental Protection of the Río de la Plata and its Maritime Boundary: Pollution Prevention and Control and Habitat Restoration. http://www.freplata.org

Gambier B, Lago DH (1995) The environment and its recent constitutional introduction. El Derecho Journal 163:727-737

Goransky R, Natale O (1996) Methodological bases to establish local regulations on water quality for human consumption. Final Report. Under-secretariat of Water Resources of the Nation, National Institute for Water Science and Technology, Buenos Aires

Instituto Argentino de la Energía 'General Mosconi'(1999) Informe de conyuntura del sector energético No. 119. Síntesis, Buenos Aires

Ingalinella AM, Stecca LM, Bachur J, Sanguinetti G, Vazquez HP (1997) Tratamiento del agua para consumo humano. In Agua: Uso y Manejo Sustentable. Seminario Internacional, Asociación de Universidades, Grupo Montevideo. Editorial Universitaria de Buenos Aires, Buenos Aires, pp 89-100

Ombudsman's Office, Association of Neighbours of La Boca, Centre for Legal and Social Studies, Co-Office of Public Defender of the City of Buenos Aires, Environment and Natural Resources Foundation, Ciudad Foundation, Foundation for Citizen Power, and Technological National University, 2002. Report, Buenos Aires

Permanent Committee to Monitor Water Quality on the South Coastal Strip of the Río de la Plata (1997). Quality of the Waters of the South Coastal Strip of the Río de la Plata (San Fernando - Magdalena) Vol. I and Institutional Vision, Vol. II, Buenos Aires

PRODIA (2000) Environmental Provincial Digest. Secretary of the Environment, 6 volumes, Buenos Aires

Tortajada C (2001) Environmental Sustainability of Water Projects. Royal Institute of Technology, Stockholm

Valls M (1999) Environmental Law. Ed Ciudad Argentina, Buenos Aires

WCDE (1987) Our Common Future. Report of the World Commission on Environment and Development. Oxford University Press, Oxford

Water Quality Management in Chile: Use of Economic Instruments

Guillermo Donoso and Oscar Melo

14.1 Introduction

Economic development has brought important improvements to the quality of life of the people in Chile. An important component of this development has been the economic activities that use water bodies as a destination for discharges. Conflicts over the use of water are becoming increasingly common. Even though the operation of markets to allocate water has been relatively successful, disputes over water quality are on the increase. This is why a national policy to manage water quality is urgently needed to avoid the social cost of inadequate quality. To put this proposal into proper context, it is necessary to explain a little more about Chile and its water management policies.

Chile is located in the south west of South America, with Argentina to the east, Peru and Bolivia to the north and the Pacific Ocean to the west and south. The country is a narrow strip of land, more than 4,200 km long. It is as narrow as 90 km and as wide as 375 km across. In 2002, the population was approximately 15.5 million, living mostly in the Metropolitan Region (39.4 percent), Region VIII (13.0 percent), and Region V (10.4 percent) (Figure 14.1). According to official figures, the urban population accounts for approximately 85 percent of the total population.

The most densely populated cities are the nation's capital Santiago (Metropolitan Region), Viña del Mar and Valparaíso (Region V) and Concepción (Region VIII). Chile's demographic growth rate is moderate, reaching 1.2 percent between 1992 and 2002 (INE 2002).

The diverse geography of the country provides a great variety of climatic conditions and a number of short river valleys running from the Andes to the Pacific Ocean. Northern Chile is arid and dry and features the Atacama Desert and a number of small river valleys. Central Chile has a temperate, Mediterranean climate with mild, wet winters and long, dry summers. In this region, the precipitation, which occurs mainly in the winter, is stored as snow and released in spring and summer when the snow melts. Southern Chile, on the other hand, is cool and wet, with rainfall, which is especially heavy in autumn and winter, throughout most of the year. Precipitation across the country ranges from near zero in the north to an annual 2000 mm in the south (Figure 14.2).

The Chilean economy is stable and has grown steadily over the last few decades. The per capita growth rate between 1984 and 1997 was among the highest in the world. In recent years, however, this rate has fallen significantly to yearly average of 2.58 percent between 1998 and 2003. The reason for the decrease is related to a drastic fall in employment explained by higher hiring costs due to discussions of a labour reform, a higher minimum wage as well as political uncertainty (Bergoeing and Morande 2002). Johnson (2001) concludes that currently Chile has a sustainable growth rate, with no risk of inflation.

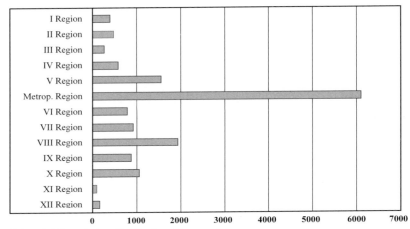

Fig. 14.1 Population distribution (thousands)

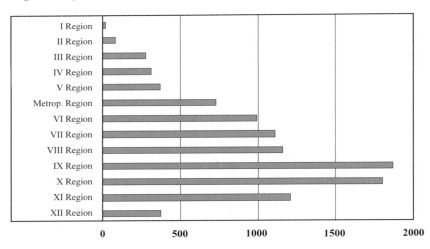

Fig. 14.2. Average rainfall (mm/year)

The legal and economic framework of the country is clearly orientated to market forces. Water management policy in Chile is worthy of note because of its innovative use of market forces to allocate water. The 1981 National Water Code designated water as national property to be used by the public, granted permanent and transferable water use rights and established water markets. The allocation of secure property rights has, in general, been beneficial to different economic sectors. Agriculture has increased its investment in irrigation technology and expanded its fruit production, while the development of drinking water plants and mining activities in areas known for their water scarcity has grown thanks to the acquisition of water use rights from agriculture.

A framework of regulations and incentives governs the use of water of each economic sector. As Peña (2001) points out, the main role of the State is orientated to defining the regulatory and institutional framework related to water resources management and, for social equity reasons, ensuring that water is available to low income families.

The State also undertakes large irrigation projects, conducts research, measures and determines the availability of water resources, regulates water use and quality, regulates services and protects and conserves water resources (Peña 2001). As Peña (2001) points out, from the perspective of the State, a positive feature is the separation of the multiple tasks carried out by the government into diverse institutions such as: water use management, General Water Directorate (DGA); water quality and environment conservation, National Commission of the Environment (CONAMA); regulation of natural monopolies (such companies that provide drinking water), Superintendence of Sanitary Services (SISS) and National Commission of Energy (CNE); and the development of irrigation infrastructure, National Irrigation Commission (CNR) and the Water Works Directorate (DOH).

The national policy of providing drinking water and sanitation services (WSS) through private regional and local water companies has been a notable success. Chile has a high level of access to drinking water and sewerage systems, comparable to those in developed countries. In 2000, 99.1 percent of the urban population and 72.3 percent of the rural population had access to drinking water. In addition, from 1990 to 2000, sewerage coverage in urban areas expanded from 86.9 percent to 93.4 percent and from 19.1 percent to 31.7 percent in rural areas (Salazar 2004).

For the last two decades, the environmental issues have become an increasingly important consideration in the legislative agenda. The first significant reference to environmental quality appears in the Constitution of 1980, which defends the right to live in a pollution-free environment. Later in 1994, Chile passed the so-called Environmental Framework Law and simultaneously created the CONAMA, whose goals are to:

- Coordinate national environmental policy with ministries;
- Supervise the national system of environmental impact assessment;
- Establish norms for environmental quality;
- Implement decontamination plans when environmental standards are not met;

Therefore, CONAMA serves as a co-ordinator of national environmental policy, whereas ministries are supposed to implement this environmental policy. The Commission has defined an environmental policy agenda for the 2002-2006 period that establishes environmental goals for each media. In the case of water quality management, it has established emission standards for discharges of wastewaters into sewerage systems, surface and groundwater, as well as enforcement policies and penalties and fines for non-compliance.

However, water quality standards for surface water have not been established for all pollutants. Several government agencies have enforced water quality management policy at several levels. The role of the DGA in water quality has been minimal. It has conducted a census of all polluters of water bodies. Other agencies responsible for regulating wastewater discharges are the Superintendence of Sanitary Services (SISS), Agricultural and Livestock Service (SAG) and the Ministry of Health.

Command-and-control instruments have been the only policy instrument used for water quality management in Chile. This strategy was heavily criticised for failing to reduce water pollution levels in the country. This is mainly due to discharges of oxygen demanding wastes, suspended solids, oil and grease, and metals from domestic and industrial sources.

In 1992, discharges of domestic and industrial wastewaters totalled approximately 43 m^3/s in the whole country. Of this amount, domestic wastewater discharges represented approximately 56 percent, and the rest were industrial wastewater discharges (SISS 1993). Point sources, including domestic and industrial wastewaters, are responsible for a discharge of 11.4 kg/s of organic waste; 60 percent of these are due to industrial wastewaters and the rest to domestic wastewaters. The contribution of nonpoint sources, mainly from agriculture, has not yet been quantified.

Water pollution in most of Chile is due mainly to a lack of proper domestic and industrial wastewater treatment. Water supply and sanitation companies are responsible for wastewater collection and treatment. From 1989 to 2002, wastewater treatment coverage expanded from 8 percent to approximately 42 percent of the population (Salazar 2004). CONAMA has set an ambitious programme to expand wastewater treatment so that 80 percent of raw domestic wastewaters will be treated by the year 2006 and 95 percent by 2010. CONAMA also aims to expand industrial wastewater treatment so that 50 percent of these discharges will be treated by the year 2006. Thus, projected increases in wastewater treatment investment should significantly improve surface water quality.

New wastewater treatment plants are being built under "build-operate-transfer" (BOT) contracts. Between 2002 and 2010, nearly $750 million will be invested in wastewater treatment plants (SISS 2002). Since WSS tariffs determine that investors receive a low-risk return of at least 7 percent on capital expenditures, companies have an incentive to invest in wastewater treatment (Verstegen 2002; Orphanópoulos 2004).

Ultimately, the WSS consumers will cover this cost through increased rates. The rate structure that SISS enforces has eliminated cross subsidies common in WSS systems and introduced an innovative programme to grant direct subsidies to poor households. As a result of current investments in wastewater treatment, surface water quality should improve. Clearly this will increase the relative importance of industrial wastewater discharges.

Water pollution is a significant problem and has major social costs. Consequently, designing and implementing a more efficient water pollution control policy is absolutely urgent and necessary. Therefore, Chile, like many other countries, has decided to reform its water quality management, adopting market instruments to introduce greater flexibility and efficiency in pollution control policies. An example of this shift is the fact that CONAMA has set a target of having a pilot programme of tradable emission permits in operation by 2005. This programme will be implemented so as to evaluate its workability and applicability to water quality management in the country.

The use of economic incentive instruments for environmental protection has gained popularity in the last few years (Eskeland and Jiménez 1992; Sánchez 1993a, 1993b), due to the limited resources for their control and the unsatisfactory results of command-and-control strategies (Bernstein 1993; Cropper and Oates 1992; Eskeland and Jiménez 1992). In developing countries, however, there are only a few examples of successful application of economic instruments in existence (for a comprehensive review of experiences in non-OECD countries, see von Amsberg 1995).

This chapter presents a policy proposal for controlling water pollution in Chile, based on the decision of the country to shift the regulatory focus from command-and-control instruments to the use of economic instruments. The focus of the proposal is to meet water quality standards (within the policy) in the most efficient way. For this purpose, the policy, including the selection of instruments, has been designed after taking the institutional set-up of the country into consideration. This requires adapting regulatory policies used in developed countries to the framework of a developing country.

First of all, the actual water quality conditions are summarised. Secondly, the main policy instruments used for controlling water pollution are described, highlighting their advantages and disadvantages and limitations and summarising results obtained from their application in several other countries. Thirdly, a policy proposal for Chile is discussed. This involves the following four phases: designating possible water uses, setting water quality standards, discharge management and education. Fourthly, a roadmap for implementing the proposed water quality management policy in Chile is presented. Lastly, some final considerations are included.

14.2 Policy Instruments for Water Quality Management

Policy instruments can be grouped into two main categories: (a) market-incentive instruments, which aim to modify the behaviour of the emitting agents by means of changing the relative prices they face and (b) command-and-control instruments, which usually establish source specific restrictions either over emissions or over the technology.

Command-and-control instruments have been, by far, the most widely used environmental policy all over the world. This approach requires authorities to define environmental quality standards and then to determine emission or technological standards to be applied to the sources. Governments must also determine enforcement policies, penalties and fines for non-compliance. The main advantage of this approach is that it gives the regulator a reasonable degree of predictability over pollution reduction levels (Bernstein 1993).

Although command-and-control strategies for pollution control have had some success in achieving pollution reductions, the approach has been criticised because it is economically inefficient and difficult to implement. The inefficiency is caused by its inflexibility, which does not allow source heterogeneity to be exploited. In other words, it does not allow the sources to take advantage of differences in their marginal costs of abatement. Even though there may be, within the command-and-control approach, some differences in requirements, for example between old and new sources or between different types of industries, there are no incentives for the sources to allocate emission reduction efforts between them efficiently after considering the relative efficiencies in pollution abatement. The result is that the desired emission reduction is not achieved at the lowest possible cost.

In addition, command-and-control instruments provide little incentive for innovation and the adoption of new technologies since, once emission or technological standards are met, there are no incentives to try for an extra effort of emission reduction. In the 1970s, almost all environmental policies relied on command-and-control instruments. There were very rare instances of economic instruments. Since the late 1980s, on the other hand, whenever a new policy is proposed, policymakers consider at least, and often select, an economic instrument. That said, almost all water quality management policies are a mixture of both, beginning as a command-and-control policy and then having economic elements added or substituted (Harrington and Morgenstern 2004).

Many countries have been adopting, or have been considering the adoption of, market instruments to introduce more flexibility and efficiency in pollution control policies (Bernstein 1993; Sánchez 1993a; von Amsberg 1995). These instruments generate the proper incentives so that polluting sources search for and adopt the most efficient compliance strategies. Despite these considerations, there are situations where the adoption of economic instruments may not generate significant cost savings over a command-and-control approach. This occurs when polluting sources present homogeneous abatement costs and under very strict environmental regulations which require significant abatement investments (Harrington and Morgenstern 2004).

Nevertheless, economic instruments, when properly applied, have a number of advantages over command-and-control instruments (OECD 1989). Their main advantages are cost effectiveness, reduced informational requirements, development of pollution control technology and being a source of revenue for the government.

One of the major disadvantages of economic instruments is that their effect on environmental quality cannot be as precisely predicted as with the command-and-control instruments, since each source has more flexibility in defining its own solution. Furthermore, regulated firms are more likely to oppose economic regulations than command-and-control because they fear they will face higher costs, despite the greater efficiency of economic instruments. Harrington and Morgenstern

(2004) find that the cost-efficiency of an economic instrument is tempered by evidence that polluting firms prefer a command-and-control instrument because of their perception that it represents lower costs for them. In all but one of the case studies conducted by the two authors, the actual or potential revenue raised by economic instruments had to be reimbursed in some way to the firms.

From the point of view of a developing country, another potential limitation of economic instruments is the high level of institutional development required to apply them, especially when using tradable permits or effluent charges (von Amsberg 1995). More specific attributes for each instrument are discussed in later sections.

In addition, it is important to acknowledge that economic instruments do not eliminate the need for regulation and enforcement by the government, despite their ability to reduce government participation and pollution regulation. In fact, it is common to find cases where economic instruments are used in conjunction with command-and-control instruments, as part of an environmental policy for pollution control (von Amsberg 1995; Harrington and Morgenstern 2004). The policy proposal developed in this paper combines command-and-control and economic instruments.

14.2.1 Taxes or Effluent Charges

With this instrument, the regulator sets the tax rate per unit of emission and the source decides how much to emit. Its main advantage is that it has the potential to be cost-effective from both a static and dynamic point of view. In a static sense, the criterion refers to the attainment of the environmental goal at the least possible cost. In a dynamic sense, the criteria relates to whether or not the instrument provides the incentive for the sources to adopt new technologies as they become available, so as to maintain inter-temporal efficiency. On the other hand, it is necessary to monitor the emissions to determine the amount of tax to be levied. Thus, when the number of sources is large, it may even be the case that monitoring costs outweigh the benefits of the policy.

In practice, taxes or emission charges have been used extensively in Europe and the United States for controlling water pollution problems. The practical cases reviewed next are summarised from Hahn (1989), Anderson et al. (1990), Bernstein (1993), Cropper and Oates (1992), Duhl (1993), and von Amsberg (1995).

France, for instance, has had an emission charge system since 1969. It is applied to all sources discharging pollutants to a water body. Other OECD countries that use emission charges for water pollution control are Australia, Germany, Italy and the Netherlands. The French system is applied to a variety of pollutants: suspended solids, dissolved solids, BOD, COD (chemical oxygen demand), organic nitrogen, ammonia, and total phosphorus. The tax is levied on all sources. Municipalities charge households on an annual basis. A nationally determined fixed rate is applied to other sources whose emissions are in some cases measured and in others only estimated.

The main goal of the system has been to raise revenues for water improvement projects, not to serve as an incentive to modify the behaviour of the sources. In fact, even though the use of taxes has been widespread, the rates have been set too low. Hahn (1989) reports that the main contribution of this system to water quality improvement has been through the financing of abatement and control activities. Its gradual introduction is one of the reasons why the system has worked well. It is important to mention that, in France, emission charges are used alongside other instruments, mainly permits.

The Netherlands also has an emission charge system for the purpose of financing water quality improvement projects and, as in France, the instrument is combined with permits. The charges depend on the volume and concentrations of pollutants, but actual discharges are only monitored at large sources. Small ones pay amounts that are unrelated to their discharges. Bernstein (1993) claims that the Dutch system has been effective and gives the right incentives to industries, such as the chemical, agricultural and bottling industries. However, the system has suffered from enforcement problems due to the existence of nonpoint sources.

The German effluent charge system, implemented in 1981, is clearly designed to provide economic incentives within a direct regulation system. The federal authorities specify water quality objectives and technological standards for the residential and industrial sectors, and charges are applied to organic material, BOD, COD, mercury, cadmium, and toxic substances for fish.

Sources are subject to a fixed discharge rate per unit. The system has a built-in mechanism that introduces tax rate reductions for sources emitting lower quantities than their emission limits. For example, if the industry complies with the emission standard, the rate is 1 percent but if it emits only 75 percent of the standard, the rate is 0.5 percent (OECD 1989).

The United States has started to implement this instrument fairly recently, mainly for the purpose of raising revenue to finance the cost of issuing discharge permits. Only some of the state programmes have an incentive objective. Currently, some 39 of the 50 states have applied an emission charge to domestic wastewater. In all cases, the charge is applied in conjunction with federal or state emission permits (Duhl 1993).

In Mexico, the environmental authority introduced an effluent charge for all sources, based on COD, suspended solids and volume. There is no charge for discharges that meet the concentration standards. The system is not effectively oriented towards emission control and, because revenues do not go to the enforcing agency, little effort has been devoted to enforcement. Therefore, little has been achieved in terms of water quality improvement (von Amsberg 1995).

Recently, four Brazilian states applied, or are in the process of applying, effluent charges. In Rio de Janeiro, the charge will be applied to all those who discharge and will be based on the volume and concentration of the discharge, including BOD and heavy metals. The fines for non-compliance, defined as a rate per unit of emission, are used to support financially the State Environmental Agency, which is the agency responsible for collecting the revenues. In the state of Sao Paulo, a charge on effluents was approved in 1993, but its implementation has been only partial. The rates are based on the average emissions of BOD and suspended solids by each sector. A revised version will allow discounts if the sources are able to prove that their cleaning processes are better than average. In the state of Parana, the system is similar to the one in Sao Paulo and has only recently started to be implemented. Finally, in the state of Minas Gerais, information is being collected in order to apply a system similar to the one applied in the other states (von Amsberg 1995).

The Colombian Water Pollution Charge Programme constitutes a valuable case study for studying the various elements for the implementation of an environmental economic instrument in developing countries, particularly an environmental charge on pollution. The charge was designed primarily as an economic incentive to reduce discharges of pollutants. However, there is evidence that this system has run into implementation problems because the institutional development required to apply the system is inadequate. Only in areas with an adequate institutional framework has this experience been successful (ECLAC 2001, 2002).

Overall, international experience indicates that effluent charges can effectively improve water quality to a degree. However, the rate must be set at a high enough level to make it a real incentive for abatement and to make the design of the appropriate institutional framework worthwhile.

14.2.2 User Fees or Tariffs

This instrument is defined as a fee charged for wastewater collection, treatment or disposal, where the cost of providing the service is charged to all users discharging wastewater into the public network of sewers (Bernstein 1993). In the case of OECD countries, there is evidence to indicate that the user fee in fact covers the handling costs (OECD 1989). Moreover, if the user fee is related in some way to the flow or quality of the discharges, it will serve as an economic incentive for reducing such discharges. Chile has implemented these user fees for residential and industrial users who discharge into sewerage systems with a water treatment plant. The charge varies and is based on the amount of water consumed. Tariffs are automatically adjusted for inflation but reviewed every five years so that the concessionaires (of sanitary services) cover operating and capital costs with at least a 7 percent return on assets and, thus, companies have the proper incentive to invest in wastewater treatment (SISS 2002; Verstegen 2002).

The impact of this policy has been a significant increase in wastewater collection and treatment coverage. From 1989 to 2002, wastewater treatment coverage expanded from 8 percent to approximately 42 percent of the population (Salazar 2004).

The user fee charged to residential users for the use of the sewerage system can be fixed, variable, or a combination of both. It can be based on the property value, the amount of water consumed, or it can simply be a fixed amount. In many countries, industrial firms are also charged for the wastewater they discharge into the network of sewers. Because of the high costs involved, direct monitoring of the quantity and quality of the discharges is usually restricted to large sources. The remaining users generally pay a fee, which may be based on the type of industry or on the amount of water consumed.

14.2.3 Subsidies

Subsidies are also used as an instrument for water pollution reduction. However, although they oblige the source to consider the effect of its discharges, and in that sense they work in the right direction, subsidies have a number of perverse incentives that could accentuate the pollution problems. The reason is that a subsidy may improve the profitability of the polluting firms and encourage new firms to become active in a particular sector. It is even possible that pollution levels end up being higher than they were in the original situation (Baumol and Oates 1988). Moreover, subsidies are not consistent with the polluter pays principle, which has been adopted by several countries.

Inspite of these problems, subsidies are widely used. In the United States, the Clean Water Act allows subsidies for planning, design and construction of wastewater treatment plants owned by the municipalities. Other countries that use this type of subsidy are Brazil, Chile, Germany, France and Italy.

Given that the tariff structure enforced in Chile has eliminated cross subsidies, the federal government has introduced an innovative programme to grant direct subsides to poor households. The water consumption subsidy is one of the few individual means-tested subsidies applied in a developing country. This programme provides incentives for poor households to limit their consumption and incentives to companies to provide service to all households (Gómez-Lobo 2001).

14.2.4 Emission Standards

This instrument consists of defining an emission limit for every source. The source can comply with its limit in any way it decides. To apply this instrument, the emissions have to be monitored in order to determine whether or not the emission limit is being satisfied. In addition, there has to be a clear system of penalties and fines for those who surpass the emission standards.

In the United States, the Federal Water Pollution Control Act (FWPCA) regulates pollutant discharges from fixed sources, such as industrial firms and wastewater treatment plants, through a system of emission standards and permits for each industry. The Environmental Protection Agency (EPA) determines water quality standards and the states establish permitting systems for the individual sources. The emission standards for the sources are set after referring to the technological standards established by the EPA. All industrial and municipal polluting sources are subject to effluent standards, according to the Clean Water Act. The standards based on the Best Practicable Technology (BPT) focus on conventional pollutants, such as BOD, dissolved oxygen, suspended solids and metals. They take into consideration the age of the equipment, processes involved, engineering aspects of abatement technologies, environmental impacts, and equilibrium between costs and benefits of emission reductions. The standards based on the Best Available Technology (BAT) are stricter and are applied to toxic pollutants. In addition to these emission standards, EPA has defined more than 115 water quality standards (Anderson et al. 1990). When evaluating the water pollution policy in the United States, Freeman (1990), concludes that, since the approval of the Clean Water Act in 1972, water quality has not improved significantly with regard to conventional pollutants.

In Mexico, federal laws set emission standards for the main industrial activities and the pollutants associated with each of them. Stricter emission limits are set for those cases in which water quality standards will not be met by using the emission standards for the industry as a whole. There are also specific standards for industrial effluents discharging into the municipal sewerage system and a system of discharge permits which insists on all existing individual sources satisfying the established standards or presenting a plan indicating how they are going to meet those standards. New sources are required to satisfy the standards when initiating operations.

In Brazil, individual emission standards and equipment requirements are negotiated case by case between the state environmental agencies and sources, becoming an essential part of construction and operation licenses. The standards and requirements can vary substantially among similar plants which are located in different places.

Argentina has also adopted emission standards. The province of Buenos Aires set regulations on waste discharges and emissions so as to reach an ambient quality this is as near the guideline level as possible. This experience has presented positive results when there is a favourable institutional framework.

The main problem associated with this instrument, by far the most widely used in water pollution control, is that it is generally not efficient or cost-effective, due to its lack of flexibility.

14.2.5 Permits and Licenses

In the United States, emission standards constitute the basis for the discharge permitting system. The permits limit the discharges of every individual source. The limitations established in the permits are of two types: those based on technology and those based on water quality. The former establishes discharge limits based on EPA regulations (discharge guidelines) for a number of industries, or based on stricter State requirements that can be applied to some types of sources. The State prepares the latter types of permits, which are based on what is considered necessary to protect the uses assigned to the water bodies being affected. The permits are issued for a certain period of time and have to be renewed periodically for companies to continue discharging. The enforcement of the permits is based on two types of monitoring: the evaluation of reports made by the own sources and spot checks made by the enforcing authority.

14.2.6 Tradable Emission Permits

Under a system of tradable emission permits, the regulatory authority determines the total amount of emissions of a given pollutant allowed in a certain region, but allows the market to spread these emissions among the sources. For this purpose, a number of permits, consistent with the previously determined total emission limit, are issued and distributed to the sources, which can trade them. The market determines the price of permits.

With a system of tradable emission permits, emissions of each individual source must be monitored, in the same way as with an emission charge system or an emission standard system.

The transaction of permits in a competitive market guarantees that the desired emissions reduction is attained at the lowest possible total cost. Those who discharge will compare the market price of the permits with their marginal abatement costs. The dischargers with low marginal abatement costs will prefer to treat their wastes rather than buy permits in the market, or if they already own permits they will sell them. Dischargers with high marginal costs of abatement will prefer to buy permits in the market and not to abate the emissions. Therefore, if there are technological differences between the sources, as expected, and the marginal costs of abatement differ, a market for permits will emerge. This market will allocate the emission reduction efforts efficiently, with those able to reduce emissions more cheaply concentrating on an emission reduction effort.

It is well known (Baumol and Oates 1988) that a tradable emission permit is equivalent to an emission charge if the tax rates are properly set. This parity is no longer valid if there is uncertainty about marginal abatement costs.

It is also true that tradable permits have a number of advantages over emission charges. To begin with, tradable permits reduce the uncertainty about the desirable global emission limit to be achieved, since the authority sets the total limit and issues the permits accordingly. What may happen, however, is that, even if the overall emission levels are the desired ones, the quality standard may be exceeded in some specific spots. This phenomenon is an aspect that has to be considered when designing a water pollution control system.

A second advantage of tradable permits over other policy instruments is their flexibility in adjusting to changes in the general economic situation. If new emission sources want to enter the market, they either have to buy permits or enter the market without emitting pollutants. Thus, the demand for permits will increase, raising the unit price of a permit. Firms that own permits will have an incentive to reduce their emissions, thus liberating permits which could be sold to newcomers or sources in expansion.

The tradable permit system also encourages sources to keep abating emissions in order to release permits that can be traded on the market. A key factor for this incentive to work is that the source must be convinced that the released permit can be traded and that the source will thus be able to benefit from the transaction. A way of assuring a firm that this will be the case is handing over some type of property right with the permit.

The distributional aspects of a tradable emission permit system depend on how the initial allocation is made. It can be done through a bidding process, or according to historic rights.

According to published accounts (Hahn 1989; Hahn and Hester 1989; Anderson et al. 1990; Tietenberg 1992), since 1980 there have been 37 programmes designed to apply tradable permits to control water pollution in the United States. Of these, 11 have been implemented, five are being implemented, six have completed their development stage, and 12 are in different discussion and study phases.

The first such programme dates back to 1981. It operates in Wisconsin for point sources discharging effluents into the Fox River. Since the programme was started, few transactions have occurred because of very strict trading restrictions. The main one is that the source must justify the need for permits, eliminating transactions based on reductions in operational costs. The duration of the permits of five years is another disincentive, since five years is too short a time for the source to recover abating investments. Transaction costs have also been high, since the discharge permits have to be modified for both plants before a deal is approved. The presence of toxic organic compounds in the discharges has also restricted trading. Some fear that toxic concentrations in the River may be on the increase.

Another example of emission trading can be found in the Dillon Reservoir in Colorado. In 1984, the state approved a phosphorous emission trading programme. The permit allows for the discharge of wastewater with a specified annual content of phosphorous. Point sources can only exceed the limit approved in 1984 by acquiring permits from other point or nonpoint sources that existed in the basin before 1984. The rate at which two point sources trade is one to one. But this increases to two to one in cases of a transaction between a point and a nonpoint source because of the greater uncertainty existing in the control of nonpoint sources. Due to the major differences in the marginal abatement costs between the sources, important savings are expected, together with an improvement in water quality.

14.3 Water Pollution Control Policy Proposal

The following pollution control policy proposal takes into account the above evidence on the efficiency of the different instruments, economic and social equity principles, such as the minimum cost of satisfying an environmental standard, and cultural principles, including the level of institutional development and previous experiences with market-oriented instruments.

The main objective of Chile's water pollution control policy is to maintain the socially desired water quality level efficiently under a second best point of view. In order to implement the water pollution control policy, it is necessary to:

- Designate possible water uses;
- Design and establish water quality standards;

- Manage discharges;
- Establish an education programme.

The policy can be based on different instruments. Their selection is based on the following criteria:

- Technical efficiency, measured as a percentage of the number of times the environmental standard is satisfied;
- Economic efficiency, defined as cost minimising;
- Equity, which implies that the cost incurred by each source must depend on its emission levels and quality;
- Operability, which is directly related to the application of the necessary control and enforcement mechanisms; and
- Acceptability, which is a measure of the degree of political acceptance.

14.3.1 Designation of Possible Water Uses

The following elements are considered when designating the use of a water body: natural water quality, current water quality, current and potential pollution sources, current and potential uses and users, abatement costs, and natural variations in water quantity and quality.

Based on the above criteria, Table 14.1 outlines a water use proposal. It should be noted that these uses are uniform across the country. Only three possible regular water uses are considered for continental surface bodies of water, along with two categories of exceptions. These categories of exceptions consider bodies of water whose natural quantity and quality does not allow their classification under a regular category and those where society is willing to pay for a higher quality. Groundwaters are classified in two categories, without exception. Estuaries and coastal waters are classified in three possible regular water categories and one category of exceptions. The exception category is designated to waters that must comply with a higher quality than required by regular categories.

Table 14.1. Proposed water classification scheme for Chile

Category	Best water use
1. Continental Surface Waters	
Regular Categories	
1C	Aquatic life propagation and maintenance (including fish), wildlife, secondary recreation, agriculture and any other usage except those indicated for Categories 1B or 1A. All continental surface waters, whose natural quantities and qualities allow it, will be classified at least in this category.
1B	Primary recreation and any other best usage specified by the 1C

	classification.
1A	Source of water supply for drinking, culinary, or food-processing purposes, and any best usage specified for Category 1C waters.
Exception Categories (studied case by case)	
1EB	Use superior to that established for Category 1A. They represent waters of outstanding quality.
1EM	Use inferior to that established for Category 1C. To be applied only to waters whose natural quantity and quality do not allow to classify them into regular categories.
2. Continental Groundwaters	
Regular Categories	
2B	Industrial and other uses, except those specified for Category 2A.
2A	Source of water supply for drinking, culinary or food-processing purposes, and any best usage specified for Category 2B.
3. Coastal Waters	
Regular Categories	
3C	Aquatic life propagation and maintenance, wildlife, secondary recreation, and any other use except those indicated for Categories 3B and 3A. All coastal waters will be classified at least in this category.
3B	Primary recreation and any other usage specified by the 3C classification.
3A	Capture of shellfish for market purposes and any other usage specified by the 3B or 3C classification.
Exception Category (studied case by case)	
3EB	Use superior to that established for Category 3A. They represent waters of outstanding quality.

14.3.2 Design and Establishment of Water Quality Standards

A water quality standard, represented by $N°$, must be designed and established for each designated water usage. Fairness requires the environmental standard associated with a certain water use to be uniform throughout the entire country.

In order to design and establish water quality standards, the regulator must at least consider the following steps: technical and economic analysis; development of scientific studies; enquiries of qualified private and public institutions; analysis of comments and observations; adequate publicity.

Based on the above, we propose that the basis for establishing different water quality standards for regular categories will be the criteria developed by the EPA. For surface waters, for instance, water quality standards for Category C will be based on chronic toxicity for aquatic species and welfare and aesthetics criteria. Standards for Category B will also include criteria for protecting primary recreation. Standards for Category A will also incorporate criteria to protect human health through water consumption (for fresh waters) and fish tissue consumption (for fresh and marine waters).

14.3.3 Discharge Management

To manage discharges adequately, the following five-point procedure is proposed: determination of the maximum amount of emissions allowed; distribution of this amount among point and nonpoint sources; management of point sources; management of nonpoint sources; and monitoring and enforcement. This procedure may need to be strengthened to adjust to emissions and to reach the desirable target.

The maximum amount of emissions permitted at a given water body, say $M°$, measured in mass load, is determined so as to ensure that the environmental standard is satisfied. It is important to consider geographic variation, the water flow's seasonal aspect and its quality in the specification of $M°$, so that all of this can be a function of the waters' self-purifying capacity. Furthermore, mathematical models of water quality must be developed and validated in order to establish $M°$ adequately and satisfy the environmental standard, $N°$.

Efficient management of pollutant discharges requires, primarily, the identification of point and nonpoint sources. Point source emissions can be quantified, while nonpoint source emissions cannot. These differences imply that efficient management of point sources is based on policy instruments that are different from those required for managing nonpoint sources efficiently.

Once all emission sources have been classified as point or nonpoint sources, it is necessary to allocate the total emissions permitted between both types of sources. Thus, a maximum permissible emissions level must be set for point sources, $M_p°$, and for nonpoint sources, $M_d°$, in such a way that the total allowable emissions, given by $M° = M_p° + M_d°$, ensures that the water quality standard, $N°$, is satisfied.

Management of point sources involves the use of a combination of an emission registry, emission standards and tradable emission permits.

All point sources must be entered in an emission register, independent of their location, type and quantity of the pollutant discharged. This register provides an inventory of emissions and facilitates monitoring and enforcement activities.

Emission standards are defined in terms of concentration (mass/volume). Emission standards for a specific pollutant are uniform across industries and throughout the country. Setting uniform emission standards prevents the "haven" effect; i.e., the emission standard does not influence the decision of the industry on a location. Uniform standards between firms, on the other hand, establish a certain degree of social equity between different installations.

Emissions discharged into a collection system are regulated by a user fee, upon which the collecting company and the polluting firm agree. The emission standard will be applied to the collecting company at its discharge point. Furthermore, the company will be held responsible for the security of the collection system (e.g. fires and explosions) and, thus, has the right to impose restrictions on the discharged quantity and quality.

The emission standards are designed and established in order to protect water quality at the point of discharge. The criteria of EPA for protecting aquatic species from acute toxicity are presented as a guideline for establishing emission standards. Emission standards defined in this way apply to lower degrees of water quality in cases characterised by a high level of uncertainty over the physical, chemical and biological behaviour of the bodies of water. This is because experts are unable to determine the maximum quantity of emissions permitted. This is particularly true for those areas surrounding discharge points.

It is important to note that, once emission standards near the discharge points are defined, water quality standards may not be met. This is to be expected because the only way to guarantee quality standards are satisfied everywhere is by setting identical emission and quality standards. But this is not feasible due to the high costs involved.

Tradable emission permits require a detailed knowledge of the natural decontamination capacity of the different "bubbles" into which the water body has been divided. It is important, therefore, to develop, calibrate and validate mathematical models that explain the quality of water, so as to establish the total emissions permitted for point sources in each bubble.

Emission permits are pollutant-specific. This facilitates compliance with environmental quality standards. In addition, emission permits must consider the seasonal aspect of water bodies' natural decontamination capacity.

Permits establish emissions in mass-load units (mass/time). The limiting factor associated with tradable emission permits is the level of uncertainty as to the behaviour of the bodies of water. This uncertainty could lead to violations of environmental standards due to a higher level of discharge than the self-purifying capacity of a water body. It is important to note, however, that the application of emission standards reduces this problem, since it sets lower expectations for the water quality achievable. The main advantage of tradable emission permits as a regulatory instrument is their flexibility, which suggests that the mechanism is efficient economically.

The permits will be allocated to different point sources for a given fee, depending on their "acquired emission rights." A fee will be determined that will partially cover the implementation costs of the regulatory policy. In saturated areas, where the environmental standard is not satisfied, the total number of permits will be limited in order to ensure the attainment of the environmental quality standard. In non-saturated areas, on the other hand, the required emission permits will be allocated and the government will keep the remaining permits and auction them off periodically.

The permits will be tradable within bubbles determined by geographical areas in which the environmental impact of the discharges depends on the total amount and is relatively independent of the location of the source. This restriction, together with the emissions standard, avoids the generation of geographical hot spots where the environmental quality standard is violated due to a concentration of emission permits.

Due to the fact that measuring individual emissions from nonpoint sources is impossible, no transactions will be permitted initially between point and nonpoint sources. Furthermore, emission permits will not expire as long as the environmental standard is respected. This is an important feature, since it provides stable and clear rules for the sources. If it became necessary to reduce emissions further, the amount of emissions allowed by each permit will be proportionally reduced. This could also be done in emergency situations, e.g. during a drought, to reduce temporarily emissions and meet the environmental standard. It should be noted that a measure of this kind ought to require an administrative order from a high authority to avoid improper use.

Violations of the annual emissions declaration, standards and permits will be sanctioned. The punishment will depend on the seriousness of each case. The quantity and quality of emissions discharged over and above the emissions permitted and the number of violations incurred previously will determine the degree of a violation. Penalties will be in the form of fines.

The optimal management of nonpoint sources is a topic that is far from settled and is still being under review. Besides an emission register, best management practices (BMPs) are considered part of the policy.

All nonpoint sources are required to be inscribed in an emissions register, irrespective of the pollutant discharged and of the source's location. This registry provides an emission inventory and facilitates monitoring and enforcement activities.

Best management practices are determined by a set of elements, such as institutional, economic and technological, which are considered as a way of reducing the environmental impact generated by nonpoint sources. In the United States the most common BMPs limit the concentration of septic systems, crop rotation, irrigation practices and pesticide and fertiliser applications (Bernstein 1993).

BMPs may also be considered a compensation mechanism between point and nonpoint sources. This mechanism will be determined on a case-by-case basis when enough information is available and both parties are interested. In this way, for example, a large point source polluter may get additional permits if it can certify a reduction at a nonpoint source. This will require the payment of compensation by the point source.

Monitoring and enforcement are an integral part of emission management from point and nonpoint sources. They merit additional discussion because they are key to a successful implementation. The enforcement mechanism is essential to ensure the effectiveness of the regulatory policy and it requires meticulous design. This is probably one of the weakest institutional frameworks and will need substantive political as well as financial support. In order to monitor emissions efficiently, the existing water quality monitoring network should include all the parameters considered in the policy proposal.

In addition, each point source will be required to present annual emissions declarations, accompanied by the required water quality analysis developed by certified laboratories. The number of water quality analyses required will depend on the specific pollutant and the source's discharge history. Finally, the enforcement mechanism includes incentives designed to modify the behaviour of the polluter, so that with a minimum monitoring effort the environmental quality standard can be satisfied.

14.3.4 Education

We consider that educating polluters, especially those from nonpoint sources, may help the long term viability of the system. Our own experience at first hand, has shown us that many agents are unaware they may be polluting. So they will not willingly enter a scheme that requires payments.

14.4 A Road Map for Water Quality Management in Chile

Although the 1980 Constitution defends the right to live in a pollution free environment, water pollution levels in Chile are generally high, mainly due to discharges of untreated wastewaters from domestic and industrial sources. This has improved in the last decades. Wastewater treatment coverage expanded from 8 percent in 1989 to approximately 42 percent in 2002 (Salazar 2004). Furthermore, CONAMA has set an ambitious programme to expand wastewater treatment so that 80 percent of raw domestic wastewaters will be treated by the year 2006 and 95 percent by 2010. CONAMA also aims to expand industrial wastewater treatment so that 50 percent of these will be treated by the year 2006. Therefore, with the current investments in wastewater treatment, surface water quality should continue to improve.

Nevertheless, it is important to point out that currently the environmental policy of the country is highly fragmented with its administration in the hands of many government agencies at several levels. Under this fragmented institutional framework, CONAMA serves as a coordinator of national environmental policy and not as the executor of the policy. In the case of water quality management, CONAMA has established emission standards for discharges of wastewaters into sewerage systems, surface and groundwater and enforcement policies, penalties and fines for non compliance. However, ambient water quality standards for surface water have not been established for all parameters.

Environmental policies also lack an effective agency dedicated to enforcing emission standards and so, in this respect, CONAMA is a relatively weak agency which relies on each ministry to monitor and enforce standards. Enforcing the current water quality management policy has been the responsibility of several government agencies at various levels, such as the DGA, SISS, SAG and the Ministry of Health, among others.

Thus, the first step required to implement an adequate water quality management policy is the strengthening of CONAMA. In this respect, CONAMA should develop and enforce regulations that implement environmental laws enacted by Congress, conduct research, set national quality and emission standards for a variety of environmental programmes and delegate to regional CONAMA offices (COREMAs) the responsibility of monitoring and enforcing compliance. Where national standards are not met, CONAMA must have the authority to issue sanctions and take other steps to assist the COREMAs in reaching the desired levels of environmental quality. The role of CONAMA and reliance upon each ministry to enforce environmental policy must be redefined so that it is allowed to actively monitor and enforce environmental policies.

Secondly, an improvement in environmental quality should be guided by the establishment of goals for ambient quality that include minimum water flows. This requires the designation of possible water uses and the design and establishment of water quality standards. In order to designate possible water uses, CONAMA must conduct a national water quality study in order to identify the significant ambient water quality issues and their location. Once water uses are designated, CONAMA must conduct a technical and economical study to design and implement ambient water quality standards. The ambient water quality standards must be based on existing environmental laws that establish that primary ambient water quality standards, which affect human health, must be uniform for all regions of the country, while secondary ambient water quality standards, designed to protect ecosystems, may vary in different geographical regions.

Once ambient water quality standards are set, saturated watersheds, which correspond to those where ambient water quality standards are not met, must be identified. For each of these saturated watersheds, a pollution reduction policy must be proposed (Plan de Prevención y Descontaminación PPDA). Chilean environmental law (Law 19.300) establishes that watersheds must be declared saturated before a pollution reduction policy can be designed and implemented (Art. 43, Law 19.300). The PPDA must determine maximum emissions allowed, the allocation among point and nonpoint sources, design and implementation of policy instruments to manage point and nonpoint sources, and monitoring and enforcement schemes.

It has been suggested that one of the PPDA be designed as a pilot programme with the water quality management proposal presented in this analysis. CONAMA proposes the application of tradable emission permits in the San Vicente Bay, as part of the water pollution control and modelling of the Concepción and San Vicente Bays in the VIII Region. This pilot programme will generate the necessary information to apply this proposal to all the country's saturated watersheds. Moreover, it will allow CONAMA to propose the proper institutional process in order effectively to implement a water quality management policy based on the use of both command-and-control and economic instruments and to monitor and enforce compliance of the ambient water quality standards.

14.5   Conclusions

Water pollution in Chile is an important problem that generates significant social costs. It is thus important to design and implement an efficient water pollution control policy that will lead to the creation of a socially acceptable environmental standard at a minimum cost. This requires the selection of cost-effective policy instruments that induce a behavioural change in the polluter. The designed regulatory policy presented in this paper exploits the complementary features between command-and-control and economic incentive instruments.

The proposed water pollution control policy makes a distinction between point and nonpoint sources. The management of point sources includes the following elements: design and establishment of water quality standards for point sources, an emissions registry, design and establishment of emission standards, the use of tradable emission permits and enforcement mechanisms. It is important to note that emission standards are uniform throughout the country in order to avoid the creation of polluters havens. In addition, emission standards set a lower standard on the water quality level in the area of discharge points. In fact, it is expected that this lower standard will violate the environmental quality standard, because setting emission standards equal to the environmental standard is not feasible due to the high costs associated. Furthermore, the use of tradable permits, combined with emission standards, avoids the perverse incentive associated with emission standards of employing deficient environmental technologies. Finally, the use of tradable permits ensures that the water quality standard will be met in a cost-effective way.

The proposal for nonpoint sources, on the other hand, specifies the design and establishment of water quality standards for nonpoint sources, an emissions registry and BMPs.

A monitoring and enforcement mechanism is proposed for both sources to ensure the effectiveness of the regulatory policy. The design of this mechanism is such that the expected cost of non-compliance is greater than the cost of compliance.

14.6 References

Anderson R, Hofmann L, Rusin M (1990) The Use of Economic Incentive Mechanisms in Environmental Management. Research Paper 051, American Petroleum Institute

Baumol W, Oates W (1988) The Theory of Environmental Policy. 2nd edn University Press, Cambridge

Bergoeing R, Morande F (2002) Growth, Employment, and Labor Taxes: Chile 1998-2001. Cuadernos de Economía 39(117): 157-174

Bernstein J (1993) Alternative Approaches to Pollution Control and Waste Management: Regulatory and Economic Instruments. In: Urban Management Program Discussion Paper No. 3. The World Bank, Washington, DC, pp 30-34

Cropper M, Oates W (1992) Environmental Economics: A Survey. Journal of Economic Literature 30(2): 675-740

Duhl J (1993) Effluent Fees: Present Practice and Future Potential. Research Paper 075, American Petroleum Institute

ECLAC (2001) Aplicación del principio contaminador-pagador en América Latina: evaluación de la efectividad ambiental y eficiencia económica de la tasa por contaminación hídrica en el sector industrial colombiano. Comisión Económica para América Latina y el Caribe, Santiago, Chile

ECLAC (2002) Aplicación del principio contaminador-pagador en América Latina. Evaluación de la efectividad ambiental y eficiencia económica de la tasa por contaminación hídrica en el sector industrial colombiano. Comisión Económica para América Latina y el Caribe, Santiago, Chile

Eskeland G, Jiménez (1992) Policy Instruments for Pollution Control in Developing Countries. The World Bank Research Observer 7(2): 145-169

Freeman AM (1990) Water Pollution Policy. In: Portney P (ed) Public Policies for Environmental Protection. Resources for the Future Press, Washington, pp 97-149

Gómez-Lobo A (2001) Incentive-Based Subsidies. Public Policy for the Private Sector Note 232, The World Bank. http://rru.worldbank.org/PapersLinks

Hahn R (1989) Economic Prescriptions for Environmental Problems: How the Patient Followed the Doctor's Orders. Journal of Economic Perspectives 3(2): 95-114

Hahn R, Hester G (1989) Marketable Permits: Lessons from Theory and Practice. Ecology Law Quarterly 16: 361-406

Harrington W, Morgenstern RD (2004) Economic Incentives versus Command-and-Control: What's the Best Approach for Solving Environmental Problems? Resources 152: 13-17

Johnson CA (2001) A Switching Regime Model for Chilean Growth. Cuadernos de Economia 38(115): 291-319

INE (2002) National Population and Housing Census, Chile

OECD (1989) Economic Instruments for Environmental Protection. Organisation for Economic Co-operation and Development, Paris

Orphanópoulos D (2004) Conceptos de Legislación Sanitaria Chilena: Tarifas Eficientes y Subsidio Específicos. In: Tortajada C, Biswas AK (eds) Precio del Agua y Participación Pública Privada en el Sector Hidráulico. Porrua, México, pp 209-238

Peña H (2001) Integrated Water Resources Management: Challenge Under the Chilean Legal-Economic Framework. In: Biswas AK, Tortajada C (eds) Integrated River Basin Management. Oxford University Press, New Delhi, pp 141-152

Salazar C (2004) El Sector Sanitario en Chile: Experiencias y Resultados. In: Tortajada C, Biswas AK (eds) Precio del Agua y Participación Pública Privada en el Sector Hidráulico. Porrua, México, pp 183-208

Sánchez JM (1993a) Instrumentos de Política para el Control de la Contaminación de Aguas Superficiales y Subterráneas. In: Antecedentes para la Elaboración de una Política de Manejo del Recurso Agua, No. 204. Centro de Estudios Públicos, Santiago, Chile

Sánchez JM (1993b) Instrumentos de Política para el Control de la Contaminación. In: Katz R, del Favero G (eds) Medio Ambiente en Desarrollo. Centro de Estudios Públicos, Santiago, Chile

SISS (1993) Memoria Anual 1992, Superintendencia de Servicios Sanitarios, Santiago, Chile

SISS (2002) Cobertura de Tratamiento de Aguas Servidas Resumen Principales Empresas. Superintencia de Servicios Sanitarios. http://www.siss.cl/default.asp?cuerpo=481

Tietenberg T (1992) Market Based Pollution Control. Paper presented at the International Conference: Soluciones Alternativas para el Problema de la Contaminación Ambiental en Santiago, Santiago, Chile (December)

Verstegen C (2002) Opportunities in the Chilean Water Sector: Report to the Netherlands Water Partnership. http://nwp.netmasters05.netmasters.nl/fulltext/fulltexthandler.cfm? fulltextevent=fulltext&objecttypeID=13-3&ID=11969&frombasket=yes

von Amsberg J (1995) Selected Experiences with the Use of Economic Instruments for Pollution Control in non-OECD Countries. In: Borregard N, Claro E, Larenas S (eds) Uso de Instrumentos Económicos en la Política Ambiental: Análisis de Casos para una Gestión Eficiente de la Contaminación en Chile, Comisión Nacional del Medio Ambiente, Santiago, Chile

Water Pollution Charges: Colombian Experience

Javier Blanco and Zulma Guzmán

15.1 Introduction

This chapter reviews the experiences of Colombia in developing economic instruments for environmental management, specifically in the area of water pollution, with a view to identifying lessons that may be applicable throughout Latin America and to other developing countries. The chapter consists of four parts. The first describes the Colombian institutional and legal context for these instruments, presenting a detailed description of the particular instrument in question (water pollution charge[1]) and of the theoretical framework within which it was developed. The second section describes the various aspects of the instrument, its implementation and results, as well as the difficulties encountered to date. The third section reviews experiences in implementing the water pollution charge, identifying both its positive and negative consequences. The last part presents conclusions for improving the instrument and recommendations for similar programmes in other countries based on the Colombian experience.

15.2 Institutional Background

15.2.1 Environmental Institutions and Legal Framework in Colombia

The economic development of Colombia has jeopardised its natural wealth[2]. As a result, over a period of more than 30 years, regulations have been introduced and institutions established to allay environmental concerns.

Colombia has one of the most advanced environmental institutional and legal frameworks in the region. The 1991 Constitution included environmental criteria as a necessary condition for economic growth.

---

[1] In November of 2003, the Ministry of Environment modified the design of the Water Pollution Charge, and at present, the modification is in a period of transition. For that reason, the chapter analyses the Water Pollution Charge with the design that operated during 1997 to 2003, and which was established by the Decree 901 of 1997.

[2] With only 0.8 percent of the land surface of the planet, Colombia has 10 percent of its global biodiversity.

Before 1991, responsibility for environmental management and monitoring was shared among various organisations, including INDERENA (a specialised subsidiary of the Ministry of Agriculture), regional autonomous corporations, whose functions include the promotion of development projects, the Ministry of Health, and municipal public utilities. As a result, functions overlapped and there was a lack of clarity regarding the scope of the mandate of each institution.

The 1991 Constitution guarantees the right to a healthy environment for the entire population and for the future generations, and assigns to the state the responsibility for securing this right[3]. Law 99 of 1993, put the principles of the Constitution into operational form and created a new framework for environmental institutions.

The framework consists of a National Environmental System (SINA), composed of various institutions with different levels of responsibility for the implementation of the policy:

1. The Ministry of the Environment[4] coordinates the system and is responsible for drawing up environmental policy, national regulations and standards, and management of national parks.
2. There are 33 regional autonomous authorities (CARs), whose main responsibility is the management of regional natural resources and protection of the environment by applying national policy guidelines and regulations.
3. Environmental agencies in each of the four largest cities[5], with responsibility for urban environmental management.
4. Five research institutions[6] to provide technical and scientific support.

Law 99 also defines the environmental responsibility of departments, municipalities and fiscal control entities.

With the 33 regional autonomous authorities and four urban environmental agencies, the environmental institutional framework in Colombia is highly decentralised. The CARs are not subsidiaries of the Ministry, but are autonomous institutions and are directed by a regional board to which the minister delegates one of approximately 10 members. Thus, the ministry's main mechanism for coordina-

---

[3] Title II. Collective and environmental rights.

[4] In 2003, the Ministry was restructured, and new functions were assigned to it, mainly the regulation and promotion of household public services (potable water, sewage and waste collection and disposal), housing and territorial planning. The new name is Ministry of Environment, Housing and Territorial Development.

[5] Environmental Management Departments are mandated for cities with over one million inhabitants. These departments are DAMA (Bogotá), DAGMA (Cali, Cauca Valley), Aburrá Valley Metropolitan Area (Medellín, Antioquia), and DADIMA (Barranquilla, Atlantic district).

[6] These are the Alexander Von Humboldt Biological Resources Research Institute, SINCHI Amazon Institute for Scientific Research, Jose Benito Vives de Andreis Institute for Marine and Coastal Research (INVEMAR), Hydrological, Meteorological and Environmental Studies Institute (IDEAM) and John Von Neumann Institute for Pacific Environmental Research.

tion is through national environmental policies, plans and regulations that the CARs implement in accordance with the local prevailing conditions.

15.2.2 Policies and Budget for Water Quality Management

Public Sector Budget for Water Quality Management

The public sector budget for water quality management comes from two main sources. The allocation from national and regional environmental authorities finances a broad range of activities, including research and public awareness, as well as support to wastewater treatment plants. The other source is the allocation by the municipalities to wastewater treatment systems as part of the budget for the water and sanitation sectors.

Water Quality Environmental Budget

The total budget of the environmental institutions in 2002 was US$245 million. This represents 0.31 percent of GDP and has remained constant for the last two years. The investment budget in 2002 was US$157 million, mainly from regional environmental authorities, illustrating the high degree of decentralisation of environmental management in Colombia.

Table 15.1. Total budget of environmental institutions in 2002 (million dollars)

Institution	Operational expenses	Debt service	Investment	Total
Ministry of Environment	12		10	22
Environmental Studies Institute (IDEAM)	8		5	13
National Environmental Fund (FONAM)			4	4
National Parks Unit	3		1	4
Regional Authorities (CARs)	60	5	137	202
Total	83	5	157	245

Source: Contraloría General de la República, 2003.

The regional authorities distribute their investment budget according to local priorities and environmental problems, taking into account the national environmental plan of the Ministry of Environment. It is difficult to calculate the total budget allocated to water quality management because these activities do not constitute a particular programme within the national plan. Although the water programme mainly includes water quality management, it also covers activities and investment related to water quantity and allocation. Furthermore, programmes such as Urban Quality and Cleaner Production may include such water quality activities as transfer of industrial technology.

Table 15.2 presents the national environmental investment, by programme, in 2002.

Table 15.2. National Environmental Investment Programme in 2002 (million dollars)

Investment programme	Central institutions	Regional authorities (CARs)	Total
Water		42.5	42.5
Biodiversity	5.8	22.9	28.7
Forests	6.3	8.1	14.4
Urban quality	0.1	26.0	26.1
Public awareness		4.5	4.5
Institutional investment	0.8	9.1	9.8
Environmental information	5.5	3.8	9.3
Green markets		0.5	0.5
Territorial planning	0.1	8.2	8.3
Participation		1.5	1.5
Cleaner production		2.4	2.4
Traditional knowledge Sustainable production systems	1.3	7.5	8.8
Total	19.9	137.1	157.0

Source: Contraloría General de la República, 2003.

The water programme received the lion's share of the US$42 million investment budget, followed by biodiversity and urban quality. Investments in the water programmes are made by the regional authorities.

Water and Sanitation Budget

Although the water and sanitation budget is implemented by both municipalities and the national government, the municipalities have the responsibility and most of the resources with which to provide safe water and sanitation. They fulfil this function through a specialised company or, in exceptional cases, directly.

The average annual budget for water and sanitation was US$260 million, almost all of it earmarked to finance the operation and investment of safe water and sanitation projects. Construction of wastewater treatment plants is marginal because the main goal is to increase the proportion of people having access to safe water and basic sanitation, currently 82 percent.

Nevertheless, Colombia has a national policy on municipal wastewater treatment that is discussed in the next section.

Table 15.3. Budget allocated to water and sanitation in 1998-2001 (million dollars)

	1998	1999	2000	2001
Municipalities	225	250	263	237
National government	1	21	12	32
Total	226	270	275	269

Source: Presentation made by the Ministry of Environment, Housing and Territorial Planning at XV Seminar on Fiscal Policy, ECLAC, 2003.

Water Quality Management: Policies

At a national level, Colombia has issued a policy document (CONPES 3177) that has set the guidelines for wastewater management for the next 10 years. According to the document, most of the 237 wastewater systems for the 1092 municipalities of the country suffer from operational problems that reduce their treatment capacities. Given the magnitude of the shortfall in water treatment in the country and the shortage of available financial resources, the document proposes criteria for prioritising the municipalities that are to construct wastewater treatment plants over the next 10 years.

The criteria include the proportion of people with access to safe drinking water and sanitation, downstream impact of the disposal, and the financial capacity of the municipality. Application of the criteria has led to the selection of 300 municipalities for the construction of wastewater treatment systems. The total cost of the systems is US$3.4 billion, while the total resources available are US$700 million. The deficit is expected to be covered by a combination of the following sources:

1. Multilateral loans;
2. Additional resources from regional environmental authorities and municipalities; and
3. Increases in water tariffs.

By the end of 2005, the national government expects to receive a loan from the World Bank of US$125 million for water and sanitation, including the financing of wastewater treatment systems. On the other hand, the government is carrying out structural reforms of the water tariff determination methodologies and regional environmental investment plans to direct resources to the plan's objectives.

15.2.3 Specific Regulatory Framework for Water Discharges

In 1974, the National Renewable Natural Resources and Environmental Protection Code (Executive Order 2811) created a legal framework of principles, rules and procedures for management of natural renewable resources and protection of the environment. Environmental charges were established by the code, but were defined exclusively as instruments for collecting fees applicable only to profit-making users (polluters) of the environment. Municipal sewage utilities were exempt from the charge.

Command-and-Control Regulation

Decree 1594 of 1984 specifically raises issues related to water pollution, defining not only environmental standards for water in accordance with their primary use, but also defining minimum standards for discharges of a variety of pollutants. In the context of this chapter, the standards established by Decree 1594 of 1984 can be considered to be the water pollution command-and-control (C&C) regulation since every water polluter was mandated to comply with the discharge standards.

On the other hand, law 99 of 1993 modified the pollution charges and created other economic and financial instruments. Consequently, after 1993, both types of regulation, command-and-control and economic instruments, had to be applied for water pollution control.

Pollution Charges

Article 42 of Law 99 of 1993 not only refers to water pollution but also to any kind of pollution of atmosphere and soil. It authorises charges for all activities (profitable and non-profitable) that discharge pollutants directly or indirectly into those resources. The regional autonomous authorities collect the charges, which become part of their financial resources. Article 42 mandates that the pollution charge has to include elements such as the valuation of the damage to public health, economic sectors, aesthetics etc. It was intended to design the pollution charge as a pigouvian tax.

In 1997, the Ministry of Environment, through Decree 901, developed a pollution charge focusing on water pollution from direct point sources. Since 1997, 33 regional agencies have levied the water pollution charge. The next section describes the main characteristics of the pollution charge, as contained in Decree 901.

15.3 Theoretical Model of Pollution Charges

15.3.1 Pigouvian Tax and Standard-and-Charge Approach

The theory of optimal environmental control of externalities prescribes that, in order to achieve a socially optimal level of abatement (pollution), the regulator must impose a tax upon the activity that generates the externality equal to the marginal net damage produced by it. This theoretical solution may seem workable despite a great amount of information demanding the implication of the valuation of all damage to health, aesthetics and economic activities: but it seems almost impossible if we consider that the marginal damage is not measured by the actual, but by the optimal, level of the externality. Finding this optimal level implies the estimation of the curve of not only the socially marginal damage costs but also the private marginal abatement costs.

For that reason, Baumol and Oates (1971, 1988) proposed an alternative approach that instead of aiming at the optimal level of pollution, helped to achieve a given environmental standard in a cost-efficient way. This approach begins by setting standards for an acceptable environment and imposing a system of charges to achieve them. An iterative process will be implemented to increase the level of the charge until the standard is achieved.

This approach is more practical to implement than the pigouvian tax because it requires significantly less information.

The standard-and-charge approach was the basis for the design of the Colombian pollution charge that was established by Decree 901 of 1997. The next section presents the main characteristics of the Colombian pollution charge.

15.3.2 Model of Pollution Charges

With Decree 901 of 1997, the Ministry of the Environment developed regulations under Article 42 in order to define the rules and procedures for implementing the water pollution charge. The charge has to be paid for the amount of pollution that a point source discharges in a river. The total cost to be paid will depend on a regional tariff of the charge and the volume of the pollutant discharged in a given period:

$$TC_{itj} = RT_{jt} * VP_{ijt} \dots \qquad \dots \qquad \dots \qquad (1)$$

where,
TC_{itj} = total cost paid by the source i during the period t for the pollutant j,
RT_{jt} = regional tariff for pollutant j in the period t, and
VP_{ijt} = volume of pollutant j that the source i discharges into the river during the period t.
The regional tariff is calculated as follows:

$$RT_{jt} = MT_j * IF_t \dots \qquad \dots \qquad \dots \qquad (2)$$

where,
RT_{jt} = regional tariff for pollutant j at the period t,
MT_j = minimum national tariff for pollutant j, and
IF_t = incremental factor at period t.

The minimal tariff (MT_j) is set by the Ministry of the Environment at a nationwide level and is indexed to inflation. In resolutions 273 of 1997, and 378 of 1998, the Ministry set the minimal tariff for BOD_5 and TSS based on a fraction of the average costs of reduction of those pollutants.

Table 15.4. Adjustments to minimum national tariff 1997-2002 in US dollars

Year	BOD ($/kg)	TSST ($/kg)	Adjustment
1997	0.017	0.007	
1998	0.020	0.009	17.68%
1999	0.024	0.010	16.70%
2000	0.026	0.011	9.23%
2001	0.028	0.012	8.75%
2002	0.030	0.013	7.65%

Source: Ministry of the Environment, 2003.

The incremental factor, IF, depends on the overall compliance of the sources to meet a specific discharge target. The target is set at the beginning of a 5-years implementation period, and is evaluated every six months. The incremental factor starts at 1 and if the target is not met, it increases in the amount of 0.5:

$$IF_0 = 1... \qquad ... \qquad ... \qquad (3)$$

$$IF_t = \begin{cases} IF_{t-1} + 0.5 & \text{if the overall discharge target is not met in period t-1} \\ IF_{t-1} & \text{if the overall discharge target is not met in period t-1} \end{cases} \quad ...\ ... \quad (4)$$

Decree 901 establishes this incremental mechanism, which is to be used for setting regional tariffs in each watershed or river sector. This factor was set by the decree, and its purpose is to induce the sources of pollution to face the charge level that will meet the target level in the least costly manner, inducing polluting parties to implement technological innovations. The gradual adjustment to the amount of the tax is an incentive to reducing pollution to the defined level as in the Baumol and Oates standard-and-charge approach.

The target is set by the board[7] of the regional autonomous authority for a five-year term. It is expressed in terms of the total amount of discharge of the pollutant, in each six-month period, to the river or a river sector, taking into account the actual and futures sources. The Ministry recommends that the process of setting the target should include a consultative process involving all the parties (polluters and community).

"The a priori determination of an environmental target is basic to the cost-effective functioning of any economic instrument, since it reflects the preferences of society in relation to environmental quality. Thus, it is important for the target to be agreed on by all sectors in-

---

[7] Members of the Board include delegates from regional environmental NGOs, industrial sectors, communities, and regional and national governments.

volved with water as a resource, including both those who produce damage through pollution and those who suffer the consequences. In this way, the costs and benefits of the decision, economic, as well as environmental and social, become a part of the decision regarding the regional target. The target also sets a benchmark for measuring the effectiveness and performance of the instrument"[8].

Once the target is set for a watershed or river section, the regional environmental authority begins to impose the charge, starting with the minimum tariff set by the Ministry. The revenue generated goes to the budget of the regional environmental authority.

The environmental authority is responsible for identifying point sources of water pollution for each river sector or for the entire watershed. Point sources are required to provide the authority with reports on discharges that constitute the basis for calculating the bills. The authority should also implement a monitoring programme to make semi-annual measurements of discharges into the watershed and compare them with agreed reduction targets.

In summary, the implementation process of the water pollution charges includes the following steps:

1. An abatement target is agreed on for the selected river or river sector;
2. Regional authority starts to charge the sources starting with the minimal national tariff;
3. Environmental authority evaluates the target on a semi-annual basis;
4. Environmental authority raises the tariff semi-annually by applying an incremental factor if the target is not met;
5. Upon achieving the regional environmental target, the rate remains stable for a five-year period; and
6. At the end of the five years, the target may be re-evaluated and changed by the members of the regional watershed community who are directly involved or affected. If the target proves to have been too ambitious, and the economic costs excessive, it should be made less restrictive. On the other hand, if the economic costs have been minimal but the environmental impact great, then the parties may agree to a more restrictive target.

The system has the virtue of allowing the community that is subject to the regulations to be actively involved with the environmental authority. Furthermore, the amount of the tax is based on measured pollution rather than subjective criteria. Thus, the system produces the minimum tax rate needed to reach the agreed decontamination target for each region, with society bearing the minimum possible cost necessary to reach the level of environmental quality desired.

---

[8] Ministerio del Medio Ambiente (1998b) El que Contamina Paga: Aguas Limpias para Colombia al Menor Costo. Implementación de las Tasas Retributivas para Contaminación Hídrica. Imprenta Nacional. Bogotá, Colombia.

15.4 Experiences of Implementation

15.4.1 Problems and Overall Issues

Implementation Process

Following the issuance of Decree 901 in April 1997, the Ministry of the Environment put in place various tools to support implementation by the regional authorities. These included an implementation manual written by experts in various relevant areas, and a capacity building programme with environmental authorities, designed to ensure that the charge principles and goals coincided with regional needs and realities.

At the end of 1997, the first year, only three of the regional environmental authorities had implemented the decree. In response, the capacity building programme was created with the aim of benefiting from the institutional strengths of those regional authorities that had made the most progress in implementing the charge (CVC, CORNARE and CARDER), so assisting other authorities in setting the programme in their regions, through in-depth capacity building sessions and control visits.

During implementation, the Ministry found that the issue related to the transparency and effective use of the revenues generated by charging was a major limitation in the process. It therefore designed a tool to provide transparency as to how the revenue is used. Information about the design of this supplementary tool, known as regional investment funds, was released at the end of 1998. The funds are intended to clarify the allocation of revenue, as well as leveraging other funding sources in order to carry out cost-effective investments in water decontamination.

The implementation of the charge was not an easy task, especially since some environmental authorities lacked credibility and political power, as well as because of the administrative infrastructure and experience needed to carry out the necessary activities efficiently.

The programme has moved forward slowly, and with significant problems. According to the Ministry, at the end of 2003, of the existing 37 environmental authorities, 33 had made progress in setting targets and implementing the instrument, while 30 have billed users in their jurisdictions for the use of the resource, and only 22 have begun forming the regional funds to handle revenue from the charge.

Implementation Aspects of the Charge

One of the main obstacles in the implementation of the charge was the treatment of sewage discharges, which can be operated directly by the municipalities or through water utilities. From the moment the decree was released, there was an extensive debate at all levels concerning who should pay the charge for the discharges from the sewage systems. On one hand, the water utilities and the former Ministry of Development argued that their main function was only to collect and transport the domestic discharges, leaving the regional authorities to bill households directly. On the other hand, the position of the Ministry of Environment was that the water utilities not only have the responsibility for the treatment of the discharges, but also have the decision-making powers and alternatives to do so. The debate ended with several changes to the law and the decree, and with an implicit agreement by the major water utilities to set the environmental goals, but not bill for the charges. According to the evaluation of the programme by the Ministry of Environment (2002), the total amount paid was only 33 percent of the amount invoiced[9].

Table 15.5. State of implementation of the water pollution charges, December 2003

Items	Regional Authorities	No. AAR
Regional environmental authorities with advanced implementation processes	Metropolitan Area, CAM, CAR, CARDER, CARDIQUE, CARSUCRE, CAS, CDMB, CODECHOCO, CORALINA, CORMACERENA, CORNARE, CORPOBOYACA, CORPOCALDAS, CORPOCHIVOR, CORPOGUAVIO, CORPONOR, CORPOURABA, CORTOLIMA, CRA, CRC, CRQ, CVC, CVS, DADIMA, DAMA, DAGMA, CORPAMAG, CORPOCESAR, CORANTIOQUIA, CORPOAMAZONIA, CORPORINOQUIA and CORPONARIÑO.	33
Authorities that have issued invoices	Metropolitan Area, CAM, CAR, CARDER, CARDIQUE, CARSUCRE, CAS, CDMB, CODECHOCO, CORALINA, CORMACERENA, CORNARE, CORPOBOYACA, CORPOCALDAS, CORPOCHIVOR, CORPOGUAVIO, CORPONOR, CORPOURABA, CORTOLIMA, CRA, CRC, CRQ, CVC, CVS, DADIMA, DAMA, DAGMA, CORPAMAG, CORPOCESAR, CORANTIOQUIA and CORPOAMAZONIA.	30*
Authorities that have not started implementation	CDA, CORPOGUAJIRA, CORPOMOJANA and CSB.	8

* 22 of the 30 Regional Authorities that have presented invoices have also established regional funds.
Source: Ministry of the Environment, Housing and Territorial Development, 2003, www.minambiente.gov.co.

---

[9] Of the total billed by the regional environmental until 2002, Col$ 66,767,729,461, the amount paid was Col$22,518,267,884.

Furthermore, the debate prevented the Potable Water Regulatory Commission from regulating the way in which the water utilities could recover the costs of the potable water tariff form households.

The environmental authorities that have made the greatest progress in implementation have had the most positive results, reducing pollution by an average of 10 to 51 percent[10].

To date, the water pollution charge in Colombia has proven effective in reducing pollution in the productive sectors. However, it has not had the same success in respect to pollution associated with sewage discharges. The technological constraints involved in controlling the latter type of pollution, high cost of implementing municipal wastewater treatment plants[11], low number of such systems currently in existence[12], and the political debate on the responsibility for household discharges have limited the effectiveness of the instrument.

In some regions (i.e. the Pacific Coast) the environmental goals of the charge were set without taking into account the specific situation of the municipalities. For example, reduction goals of 50 percent were set for domestic discharges from municipalities in which the majority of the population is poor and where the scarce resources of the municipalities were focused principally in increasing the low coverage of the sewage system. Furthermore, the flow of the river that received the discharges was sufficient to dilute the pollution to an insignificant extent.

Therefore, municipalities that have been billed are incurring growing debts to the environmental authorities, a situation that, in the long run, could create a serious budgetary problem for the authorities. Municipal entities account for 70 of the wastewater discharges at the national level. In the great majority of the cases, they have no cleansing or treatment systems. Failure to pay the environmental tax for their discharges merely aggravates the situation.

Despite this, the environmental authorities have billed water and sewage utilities, though they have not collected. They have taken legal action to enforce the environmental legislation, as in the case of urban regional authority DADIMA, which was forced to seize the Barranquilla Water, Sewerage and Street Cleaning Company (also known as Triple A). The company then agreed to pay its approximately 2.5 billion pesos debt over a period of one year, in addition to monthly payments for household sewage discharged since October 2001. Similar cases have occurred elsewhere. Medellín Public Utilities (Empresas Públicas de Medellín) is one example. This utility only began paying for discharges as of

---

[10] CORNARE has reported 31 percent BOD reduction and 47 percent TSS reduction; CDMB has found 10 percent BOD reduction and 69 percent TSS reduction; and CVC has reported more than 46 percent BOD reduction and 36 percent TSS reduction. These data were estimated by the authors based on the latest data reported by the Ministry of the Environment in 2003.

[11] Investment needed by this sector in Colombia is estimated at US$2.175 billion.

[12] 95 percent of the municipalities in Colombia release wastewater without any treatment.

2000, but the underlying problem, one of gaps in the law, was not solved. Thus, the more the charge is billed; the worse becomes the problem of debt, and the more other sectors become involved. In this connection, the National Industrial Association has expressed concern about the inequality of implementation, since municipalities are neither reducing their discharges nor paying the charge, while industrial users that discharge in the same river sector, comply with regulations and investing in cleaning-up are having to pay continuous increases in the regional factor. These increases are due to a failure to reach target levels, a result, specifically of the fact that municipalities, which are responsible for 70 percent of the discharges, are not taking any measures to mitigate them.

In addition, the failure to impose the charge all across the regions of the country creates inequities and competitive disadvantages for firms situated within the jurisdiction of environmental authorities that are actually collecting the tax. These firms must either pay fees or reduce their discharges, while firms not subject to the regulations are free of additional costs. The problem has been addressed by the Comptroller-General of the Republic, who has stressed the failure of environmental authorities to enforce the law, both with respect to collecting the charge and in the sense of enforcing the standards imposed by Decree 1594 of 1984.

Much of the debate is focused on the lack of credibility of government institutions, which are considered corrupt, inefficient and bureaucratic. The private sector opposes channelling its revenues into the general budget of the regional authorities. They believe that such a procedure will not ensure that the funds will be used to solve specific regional environmental problems. With this in mind, some of the regional authorities, under guidance from the Ministry, have made agreements with municipalities for the management of regional investment funds to ensure the design of abatement measures that provide for cost-effective treatment systems.

Impact on Households

The institutional process involved in dealing with the impact of the water pollution charge on households, however, is highly complex, involving the Ministry of the Environment, the Ministry of Economic Development, the Ministry of Health, the Commission for the Regulation of Drinking Water and Basic Sanitation, the National Planning Department, and the Office of the Superintendent of Household Utilities. Moreover, the charge is seen as an additional cost which, when transferred to users, will have harmful effects on the lower socio-economic population, which is already under economic strain. This segment of the population has already suffered, over the last two years, from the economic impact caused by the reduction of subsidies to potable water tariffs. Before implementing Decree 901 of 1997, the Ministry of the Environment analysed the impact of the tax on household users of sewage services. It concluded that the monthly cost borne by a family would be only 0.2 percent of the monthly minimum wage (approximately 600 pesos in 2002 terms).

The Ministry projected payments by municipalities and utilities in each region, based on population or number of users, the amount of pollution being discharged into the environment, and the reductions targeted for the respective regions. The analysis was based on two alternatives from which municipalities could choose: (a) paying an environmental tax to environmental authorities on an ongoing basis; or (b) investing in pollution reducing measures that are less costly than paying the tax.

Table 15.6 shows the results of studies carried out by the ministries of the Environment and of Development (CONPES 3177). This indicates that the impact of the charge itself will be on average 6 percent of the total potable water bill. However, the expected impact of including the costs of the wastewater treatment system and the programme of reducing the subsidies will increase the tariff by as much as 200 percent.

The policy document, CONPES 3177, concludes that the following measures are necessary:

1. Modification of the pollution charge in order to streamline its application in the most polluted rivers;
2. Implementation of a government plan for wastewater treatment to direct regional and national resources to the municipalities that generate major pollution problems. The other municipalities should focus on achieving a minimum coverage of the sewage system;
3. Establishment of the target for the water pollution charge with the municipalities in accordance with a feasible plan for wastewater treatment;
4. Evaluation of the possibility of adding others pollutants to the charge; and
5. Modification of the water pollution charge to reflect the above recommendations.

After the release of the CONPES document, the Ministry launched a process for the modification of Decree 901 that ended in November 2003, with the following changes:

1. Direct relationship of the charge as an instrument for planning water resources before implementation of the charge, environmental authority has to elaborate a water management plan that assesses the water quality and compares it with the quality standards according with its main uses, the plan being the main input for establishing environmental goals;
2. Standardisation of the procedure for establishing environmental goals in order to guarantee the participation of the sources and the community;
3. Modification of the incremental factor, including increment caps and setting the increments in proportion to the achievement of the goal;
4. Directives for setting environmental goals related to municipal discharges according to a wastewater treatment plan; and
5. Possibility of establishing tradable individual goals for major polluters.

Since, at the moment, the authorities are in a transition period for implementation of the modifications, it is not possible to evaluate the overall performance.

Table 15.6. Estimated impact of the water pollution charge on potable water tariff

Population strata (1 poorest – 6 richest)	Impact of subsidy reduction	Impact of pollution charge in present tariff	Impact of maximum pollution charge in future tariff	Impact of including investment in wastewater treatment
1	60.9%	4.2%	13.3%	51.3%
2	48.6%	2.8%	9.8%	38.3%
3	45.1%	2.2%	7.7%	44.4%
4	24.1%	1.7%	7.2%	43.8%
5	9.0%	1.3%	6.2%	32.5%
6	1.7%	1.2%	6.0%	23.3%
Average	33.9%	2.1%	7.5%	39.6%

Source: Departamento Nacional de Planeación, Ministerio del Medio Ambiente, Ministerio de Desarrollo (2002) Acciones Prioritarias y Lineamientos para la Formulación del Plan Nacional de Manejo de Aguas Residuales. Documentos Consejo Nacional de Política Económica y Social - CONPES 3177, Bogotá, Colombia.

15.4.2 Positive Aspects and Successful Cases

The former Director of the Economic Analysis Office of the Ministry of the Environment, made the following observations in terms of evaluation of the environmental tax at an event hosted by the World Bank as part of the release of the book "Greening Industries", on November 22, 2000[13].

"Our group has been directly involved in the design and implementation of environmental taxes with officials from China, the Philippines, Indonesia, France, Canada and Brazil during the last seven years, and we have had the opportunity to see where, in practice, there are problems related to the implementation of taxes.

After having evaluated the Colombian tax, our conclusion, comparing this with the experience in other countries, is that the Colombian mechanism represents the state of the art in the design and implementation of taxes. It is a system based on cost-effectiveness and on minimizing costs. It is a mechanism notable for its simplicity and transparency, and for the fact that is objective and gradual. It also stands out for the way in which it actually incorporates the regional community in the setting of cleanup goals that reflect local preferences. It is based on a fairly practical system and method, and empirical evidence indicates that the mechanism can be implemented with existing institutional resources to reduce pollution in a very cost-effective manner."

This conclusion appears in the case analysis made by the CORNARE regional authority, in Greening Industry, a book published by the World Bank in 2000. When the regulatory decree establishing the water pollution charge was first issued, the regional environmental authority of Antioquia (CORNARE) was 15

---

[13] Thomas Black intervention during World Bank - NIPR team Public Seminar related to its report Greening Industry on November 22, 1999. http://www.worldbank.org/nipr/greening/webcast.htm.

years old. CORNARE pioneered the implementation of the charge with assistance from staff of the Ministry of Economic Analysis.

Unlike most of the autonomous entities of the country, CORNARE had a significant amount of information on the situation and quality of the most important water resources within its jurisdiction, and knew the entities responsible for the major discharges in its watersheds. Using this information, as well as statements by the entities themselves, as provided for in the regulatory scheme, baselines were established for each of eight watersheds, based on discharges measured according to the parameters defined in the regulations.

Discharge levels for the various parameters were specified for specific discharge points within the watersheds. CORNARE had data on the location of polluters, classified by economic sector and sub-sector. This made the process of agreeing on regional clean-up targets clearer and the results more precise. Moreover, there had already been meaningful dialogue with unions within these subsectors, leading to agreements on cleaner production[14]. This was highly significant, since these groups represented over 90 percent of identified water pollution sources in the region's productive sector.

In the case of municipalities and utilities serving households, the statements received showed these entities contributing approximately 70 percent of the volume of BOD_5 and TSS.

The governing board of CORNARE approved the first six months of charges, covering April 1 to September 30, 1997, and implementation has had significant positive results in a number of the eight target watersheds. For instance, after two years of implementation in the Río Claro – Cocorná Sur watershed, TSS pollution had fallen by 84.95 percent, and BOD_5 pollution had dropped by 40.42 percent. This watershed includes cement plants and oil industry sites. The Río Negro watershed, one of the most polluted in Colombia, showed reductions of 33.81 percent in TSS and of 33.56 percent in BOD_5[15].

CORNARE has also been successful in the area of collection, having collected 57 percent of the total billed. This is far higher than in other parts of the country. Indeed, all of the industrial entities that signed cleaner production agreements with CORNARE[16] have made their water pollution charge payments in a timely fashion.

One key point for the success of CORNARE success is the transparency with which the collected funds have been managed. From the start, CORNARE determined that 50 percent of the money would be used to co-finance projects designed to deal with municipal pollution, 30 percent for investment in industrial reengi-

---

[14] Industrial entities associated with CEO (Corporación Empresarial del Oriente), flower growers (*Asocolflores antioquia*), pork producers (ACP and others), agave farmers (Asdefique and others), and beekeepers (Fenavi and others).

[15] Castro L.F., and Castro R. (2002) Tasas Retributivas por Vertimientos Puntuales. Evaluación Nacional. Ministerio del Medio Ambiente. Bogotá, Colombia.

[16] Members of Industrial Corporation of the East of Antioquia (Corporación Empresarial del Oriente Antioqueño).

neering and cleaner production, 10 percent on research for environmental science and technology, environmental education, and dissemination of information about the environmental tax, and an impressively low 10 percent as operating expenses. A regional clean-up fund was created to handle the revenue, following guidelines provided by the Ministry of the Environment for that purpose.

Table 15.7. CORNARE Collection of invoiced water pollution charge in Colombian pesos

Sector	Total billings 1992–2001 (Col $/semester)	Payments 1992–2001 (Col $/semester)	Debts 1992–2001 (Col $/semester)	% Collected
Household	2.935.587.307	1.221.806.022	1.713.781.285	58
Industrial	944.108.990	451.643.589	492.465.401	52
Agricultural sector	26.590.765	12.404.094	14.186.672	53
Total	3.906.287.063	1.685.853.705	2.220.433.358	57

Source: Estimation of the authors, based on data from the Ministry of the Environment, 2002.

The Economic Commission for Latin America and the Caribbean, ECLAC[17], has also been involved in analyses of consequences of the water pollution charge, conducting an assessment of the impact of the charge on the industrial sector in Colombia after four years of implementation. The study dealt with environmental effectiveness and economic efficiency in the jurisdictions of environmental authorities CVC, CORNARE and DADIMA, examining the two following scenarios:

1. Regulated entities have met standards on discharges, investing in costly treatment plants, as in the cases of CVC and CORNARE; and
2. Regulated entities have low levels of compliance with standards, not investing in treatment plants, as in the case of DADIMA.

The study concluded that in both cases, the tax produced significant reductions, in addition to those already produced by existing treatment plants, and more rapidly than in previous years, as shown in Figure 15.1.

In addition, ECLAC evaluated the cost of compliance for regulated enterprises, examining two individual companies, Monómeros Colombo Venezolanos S.A. and Canteras Yarumal, as well as a third company, Cultivadores de Caña de Colombia (ASOCAÑA), in the sugarcane sub-sector. The results showed that in addition to a reduction in pollution compared with the previous system, companies had incentives to create cleaner production processes, thus leading, in some cases, to a rise in productivity.

[17] Castro L.F., Caicedo J.C., Jaramillo A and Morera L. (2001) Aplicación del principio contaminador-pagador en América Latina: evaluación de la efectividad ambiental y eficiencia económica de la tasa por contaminación hídrica en el sector industrial colombiano. Serie Medio Ambiente y Desarrollo, Economic Comisión for Latin America and the Caribean, ECLAC, United Nations, Santiago de Chile, Chile.

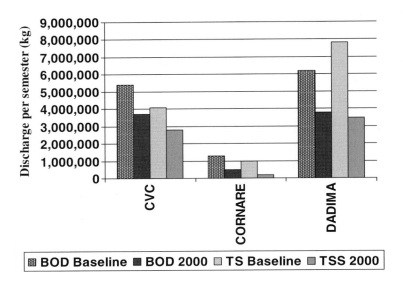

Fig. 15.1 Industrial discharges before and after the environmental tax (Source: Castro et al., 2002)

Though the sugar industry already had treatment plants for its wastes, additional reductions of 24 percent (BOD) and 65 percent (TSS) followed implementation of the charge. In the cases of the individual companies studied, approximately 90 percent reductions were achieved within a year of implementation, even when production was increasing. Such results had not been achieved previously, despite a policy that imposed fines and even shut down companies.

According to the ECLAC study, the charge, as a source of funds for the environmental authorities, gives these entities budget stability independent of the central government, guaranteeing that there will be money for new projects, and ensuring continuity of the programmes that they administer, supervise and monitor in water resources.

To summarise, the most significant benefits of the water pollution charge system, according to the various studies and analyses of its application in Colombia, are:

1. Updating of inventories of users that discharge directly or indirectly into water bodies;
2. Updating of information on the state of water resources in terms of organic pollution and suspended solids;
3. Generation of financial resources for investment in water quality management; and
4. Linking environmental instruments with quantitative targets and periodically evaluating performance.

In addition, documents, in which companies and municipalities describe the discharges they generate, are now taken into consideration. This information is of value to the management, and can be a factor in inducing regulatory entities to fulfil their environmental responsibilities. Other notable effects are:

1. In certain regions, closer and better relations have been established between environmental authorities and those responsible for discharges;
2. Obligation to pay the charge has made users more aware of environmental issues; and
3. In some regions of the country, industrial pollution has been reduced.

15.4.3 Empirical Evaluation

The following section presents two statistical analyses in order to evaluate the efficiency of the Colombian Water Pollution Charge (CWPC) in reducing discharges compared with the command-and-control regulation, and to identify the main factors that explain the discharges reduction within the implementation of the charge.

Database Characteristics

The database that was used for this chapter was based on the information collected by the Ministry of Environment Evaluation Study in 2002. The database included information on 32 regional authorities about discharges of point sources of BOD and TSS, during the periods in which the authority had monitored the sources and implemented the CWPC. It is important to note that the regional authorities do not implement the CWPC at the same time, but within different periods depending on their technical capacity and regional political acceptance of the instrument. For that reason, the data of 12 regional authorities that have reported two or more periods of implementation of the CWPC, in 2002, were analysed. We considered that one period of implementation does not give sufficient basis for a conclusion on the behaviour of the sources towards the water pollution charge.

The database contains 12,813 observations of 1,623 point sources, and each observation includes the following variables: name of the source, regional authority, river, river sector, year, semester, BOD discharge (kg) in the semester, TSS discharge (kg) in the semester, BOD invoiced in the semester, TSS invoiced in the semester, total invoiced in the semester, total paid in the semester, BOD tariff (calculated by dividing BOD invoiced by BOD discharge) and TSS Tariff (estimated by dividing TSS invoiced by TSS discharge).

Overall Efficiency of the Pollution Charge (Command-and-Control vs Pollution Charge)

To evaluate the efficiency of the water pollution charge compared with the command-and-control regulation, it is necessary to have observations of the discharges of sources under both systems. In the database only four regional authorities reported data of the discharges of sources before the water pollution charge was implemented.

Table 15.8. CAR * CONTRDBO Cross-tabulation: Observation count

Regional Authorities	Regulation		
	BOD C&C	BOD CWPC	Total
CDMB	444	482	926
CORNARE	588	1282	1870
CORTOLIMA	527	285	812
CVC	1,253	1,303	2,556
Total	2,812	3,352	6,164

A total of 6,164 observations, 2,812 of sources in periods only with command-and-control, and 3,352 in periods with water pollution charges, were analysed.

Figure 15.2 shows the means of the BOD discharges with CWPC and with command-and-control for the four regional authorities. As it can be observed, except for CONARE, there are no discernable differences in the mean levels of discharges of the sources between the periods with command-and-control regulation and with water pollution charge.

To formally evaluate the hypothesis that the water pollution charge produces a reduction of discharges, a T-test of difference of means in the variable of BOD and TSS discharges was applied to each observation of each regional authority. A summary of results is shown in the Table 15.9.

The only regional authority that presents a statistical significant reduction of discharges both in BOD and TSS as a result of applying the water pollution charge is CORNARE. For CVC and CDMB, the differences are not statistically significant. For CORTOLIMA, and partially for CDMB, the application of the charge produces a statistically significant increment on the discharges, compared with the command-and-control regulation.

These contradictory results can be explained partially by the fact that, with the implementation of the pollution charge, the authorities have improved their monitoring of the discharges of point sources. This behaviour was one of the conclusions of the 2002 evaluation study by the Ministry. The programme creates the incentive for regional authorities to update and improve their discharge data for the sources, as the revenues from the charges collected will be added to their budget. In the same way, the programme requires that the institutions responsible for point sources continuously monitor their discharges, and then report them to the authority.

On the other hand, CORNARE has one of the best national discharge monitoring programmes, so it is reasonable to think that the implementation of the water pollution charge will not create a difference in its monitored data. In the case of CORNARE, the charge produced a reduction of discharges by more than 50 percent compared with the level of discharges when command-and-control alone was applied.

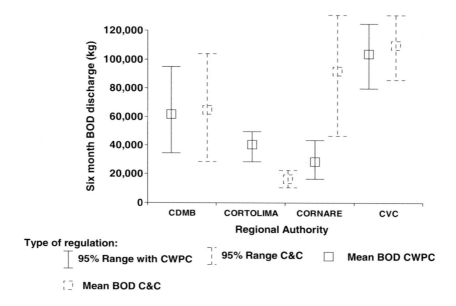

Fig. 15.2. Pollution charge vs command-and-control discharges of BOD

In conclusion, the efficiency of the CWPC with respect to the command-and-control regulation could only be proved in one of the four regional authorities that reported complete data. A possible explanation of the results in the other authorities is the incentive of the programme to improve the monitoring data sets and consequently to increase the discharges reported for the point sources.

The former conclusion does not imply that in the periods when the pollution charge is implemented, it produces discharge reduction as observed in various studies (ECLAC, MMA, WB, etc.). In fact, that is precisely the case observed in the database. The following figure shows the average reduction of point source discharges in the BOD during the periods when the pollution charge was implemented. It also shows that the CWPC generates discharge reductions during its implementation period.

The next section constructs a model to determine the main elements that explains the discharge reductions

Table 15.9. T-Test for equality of means (CWPC and C&C)

Parameters	T	Df	Sig. (2-tailed)	Mean difference	Standard error difference	95% Confidence interval of the difference	
						Lower	Upper
CORNARE							
BOD Discharge	4.149	1835	.000	15826.62	3814.16	8346.05	23307.19
TSS Discharge	5.510	1824	.000	17236.87	3128.04	11101.95	23371.80
CVC							
BOD Discharge	-.524	2550	.600	-5817.52	11103.56	-27590.44	15955.40
TSS Discharge	-.075	2521	.940	-879.72	11754.34	-23928.88	22169.43
CDMB							
BOD Discharge	-1.787	924	.074	-190271.6	106469.1	-399220.91	18677.69
TSS Discharge	-3.231	805	.001	-49696.83	15379.74	-79885.96	-19507.69
CORTOLIMA						Lower	Upper
BOD Discharge	-3.231	805	.001	-49696.83	15379.74	-79885.96	-19507.69
TSS Discharge	-3.570	802	.000	-63309.60	17733.14	-98118.45	-28500.75

Source: Estimations of the author.

Table 15.10. Discharge reduction in CORNARE water pollution, charge vs command-and-control

		N	Mean	Std. Deviation	Std. Error Mean
BOD Discharge	BOD with C&C	555	32392,8864	107192,59606	4550,07230
	BOD with WPC	1282	16566,2619	55698,61226	1555,60870
TSS Discharge	SST with C&C	556	30556,7175	87016,93284	3690,33996
	SST with WPC	1270	13319,8382	46119,35147	1294,14021

Source: Estimation by the authors.

Elements Explaining Pollution Abatement within the Pollution Charge Programme

In order to explain the discharge reduction, we aggregated the observations of the database to calculate the following variables for each point source:

- BOD discharge reduction during the period that the CWPC was implemented;
- TSS discharge reduction during the period that the CWPC was implemented;
- Number of periods, semesters when the charge was invoiced to the source;
- Final BOD tariff, the last tariff for BOD discharges invoiced to the source;
- Final TSS tariff, the last tariff for TSS discharges invoiced to the source;
- BOD tariff increment, during the period when the CWPC was implemented;
- TSS tariff increment, during the period when the CWPC was implemented;
- Total invoiced, to the source during the semesters when the CPWC was implemented;
- Total paid, total value paid by the source during the semesters when the CPWC was implemented;
- Type of discharge;
- Average BOD discharge per semester; and
- Average TSS discharge per semester.

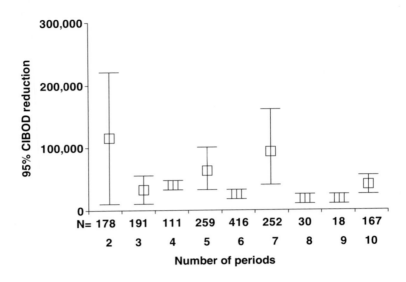

Fig. 15.3. Mean of BOD discharge reduction (kg/semester) during the implementation of CWPC

Table 15.11 presents descriptive statistics of the variables.

Table 15.11. Descriptive statistics

	No.	Minimum	Maximum	Mean	Std. Deviation
BOD discharge reduction	1622	.00	9900000.00	37278.35	327552.760
TSS discharge reduction	1622	.00	32862978.54	55995.21	866083.627
Number of periods	1622	2	10	5.55	2.24
Final BOD Tariff	1483	8.00	398.08	171.39	101.26
BOD Tariff increment	1590	.00	465.13	107.75	102.66
Final TSS Tariff	1492	2.42	480.11	79.08	47.63
TSS Tariff increment	1584	.00	460.21	51.34	49.05
Total paid	1622	.00	7924115628.00	14634308.63	226116805.33
Total invoiced	1617	1543.06	10714983198.00	42834846.83	392955908.31
Average BOD discharge	1389	100.61	36028203.83	94849.70	1186309.28
Average TSS discharge	1402	100.08	30968755.91	97474.22	1121644.86
Valid No. (listwise)	1250				

The number of sources in the database with complete information for all variables was 1,250. The average reductions per source of BOD and TSS discharges were 37 tons/semester and 55 tons/semester respectively. The average period during which the sources were facing the pollution charge was 5.5 semesters. The amount paid is 34 percent of the invoice, consistent with the findings of Castro et al., 2002, and Castro and Castro 2002.

The variables of BOD and TSS discharges, as well as the amounts invoiced and paid were transformed by applying natural logarithms (LN) in order to standardise the size of the source. The models tested were constructed in a way to avoid colinearity of the independent variables, especially with the variables of discharges, invoiced and tariff.

A stepwise method for constructing the best model was used. This method enters the variables into the model taking into account its effect on the change of the significance (sig F statistic) of the overall model. A variable is entered if it increases the significance of the model by at least 5 percent or its exclusion does not decrease it by 10 percent.

The model with the best R2 was the following:

LN BOD reduction = constant + B1 (LN (Total Paid)) + B2 (Number of Periods) + B3 (BOD final Tariff) + B4 (BOD tariff increment)

LN TSS reduction = constant + B1 (LN (Total Paid)) + B2 (Number of Periods) + B3 (TSS final Tariff) + B4 (TSS tariff increment)

Table 15.12 and Table 15.13 show the overall results of the models.

Table 15.12. Model summary

Model	R	R Square	Adjusted R Square	Std. Error of the Estimate	Durbin-Watson
1	.807	.651	.649	2.04028	1.471
2	.757	.573	.571	2.30828	1.489

Table 15.13. ANOVA

Model		Sum of Squares	Df	Mean Square	F	Sig.
1	Regression	4836.822	4	1209.206	290.482	.000
	Residual	2589.230	622	4.163		
	Total	7426.052	626			
2	Regression	4505.481	4	1126.370	211.400	.000
	Residual	3351.404	629	5.328		
	Total	7856.885	633			

1 Predictors: (Constant), LN Total Paid, BOD final Tariff, BOD Tariff Increment, No. of Periods. Dependent Variable: LNREDBOD.
2 Predictors: (Constant), LN Total Paid, TSS final Tariff, TSS Tariff Increment, No. of Pe-

riods. Dependent Variable: LNREDTSS.

Both models have good performance statistics, adjusted R square of 65% and 57% respectively, and significance of F statistic at less than 1% error.

With respect to the predictor variables, the following table presents the summary of the coefficients and individual test of significance.

Table 15.14. Coefficients

		Unstandardised Coefficients		Standardized Coefficients	T	Sig.
1	(Constant)	-3.288	.605		-5.436	.000
	LN Total Paid	.955	.032	.735	30.072	.000
	BOD final Tariff	-2.425E-02	.003	-.842	-9.450	.000
	BOD Tariff Increment	2.488E-02	.003	.874	8.248	.000
	No. of Periods	-.229	.056	-.164	-4.123	.000
2	(Constant)	-1.414	.635		-2.226	.026
	LN Total Paid	.900	.035	.675	25.362	.000
	TSS final Tariff	-6.454E-02	.006	-1.090	-10.209	.000
	TSS Tariff Increment	6.492E-02	.007	1.113	9.396	.000
	No. of Periods	-.332	.054	-.232	-6.183	.000

1   Dependent Variable: LNREDBOD.
2   Dependent Variable: LNREDTSS.

All variables are significant to the model at less than 1 percent error level. Total paid and tariff increments variables have the expected coefficient sign. That is, a greater increment in tariff produces more discharge reduction, as well as a greater amount paid by the source produces more discharge reduction. On the other hand, the variables "number of periods" and "tariff levels" have negative signs, indicating an inverse relationship between them and the discharge reduction.

In conclusion, the reduction of discharge within the WPC is explained mainly by the increment of the tariff of the pollutants and the enforcement of the charge, reflected by the amount that the sources have to pay for their discharges. The absolute level of the tariff, as well as the number of periods during which the WPC is implemented, does not have a direct relationship with the discharge reductions. Furthermore, it is possible that high levels of the tariff could result in low reductions because the sources could choose not to pay the charge. The behaviour of the variable "number of periods" could reflect the fact that the most significant reductions will be achieve at the beginning of the programme.

15.5 Conclusions and Recommendations

15.5.1 Conclusions

It is widely recognised that it is more economically efficient for society to tax "negative" activities, such as pollution, than "positive" ones such as work or savings. Hence, there is a three-way benefit in taxing pollution: it diminishes harm to the environment, generates additional revenue for environmental management and reduces taxation on activity that is beneficial to the society.

Thus, in theory, taxing pollution creates an incentive for environmentally favourable behaviour, promotes a proper valuation of natural resources, and encourages their efficient allocation and use. As a result, the environment is treated in a manner that ensures efficient economic allocation of resources, thus internalising any damage derived from production.

An accurate valuation of pollution, for example, implies that the marginal costs of reduction (or, viewed from another perspective, the marginal benefits of polluting) are equal to the marginal costs of damage to the natural resources involved, in the optimal scenario. Environmental goods and services, however, are not tradable, and the information needed to determine the marginal cost of harm done is generally unavailable. Hence, economic instruments are used – ones that, despite a lack of important information, do permit polluters to determine the most appropriate form of reaching an established target, or of making their marginal cost for decontamination equal to the level of the tax that has been set on the amount and type of pollution in question. The use of economic instruments in environmental policy is not new. Nevertheless, there is a large gap between the theory on which they are based and the actual implementation of the instruments. According to the Organisation for Economic Cooperation and Development (OECD 1994), though these instruments are the type most commonly used, they have been problematic in practice and have failed to create the incentive levels needed to attain environmental goals, mainly because the level of the charge has been set too low, and because they are perceived to be financial instruments or sources of revenue.

The situation in Colombia is different. The instrument here was designed primarily as an economic incentive to reduce releases of polluting agents into water bodies. While this is its principal strength, it suffers from other problems. These relate primarily to implementation rather than design, except for the exponential nature of the regional multiplier. The environmental charge is, in theory, an ideal instrument for pollution control, but in practice its application raises a variety of problems and its design fails to take account of such important elements as:

- Time needed to design, build and implement technological solutions to mitigate releases of pollutants;
- Low level of municipal resources available to build treatment plants needed in a timely fashion;

- Delays in bringing complete sanitation services to many municipalities, making it impossible to solve the problem of waste releases into water bodies, even with the construction of treatment plants (which would be underutilised because of the incomplete coverage of the sewage collection system);
- Elimination of subsidies to utilities leads to social opposition to the tax;
- Vagueness of regulations governing transfer of utilities of the tax to users has caused resistance by waste treatment companies, which believe that they may be forced to pay the charges out of their own pockets;
- Lack of knowledge about environmental issues, especially in terms of the basic concepts of the instrument as designed, which has caused chaos when the time comes to negotiate clean-up targets, sometimes leading to targets that are entirely unrealistic;
- Low level of experience and technical capacity of the environmental authorities, reflected their inability to explain to the regulated parties the rationale for the tax, and their lack of the credibility and capacities needed to negotiate agreements;
- Lack of information on releases of pollutants into water bodies, which means environmental authorities cannot bill and collect the charges;
- Problems environmental authorities have had in carrying out the auditing and monitoring required in order to confirm information that appears in the statements provided by entities responsible for discharges; and
- Exponential increase of the regional multiplier and, consequently, of the tax, due to failure to meet targets, a situation that has led to widespread resistance to the charge.

For the above reasons, experiences that have been perceived as successful are those involving environmental authorities that were already engaged in managing discharges by economic agents in their jurisdictions before the advent of the economic instrument, entities that had basic information on discharges into their water bodies. They have been free of most of the problems that have occurred in the rest of the country, thus managing to obtain compliance with the targets, and preventing sharp increases in the regional multiplier.

Studies analysing benefits of the water pollution charge programme have pointed to solutions to the problems of information and management essential to successful implementation, e.g., (a) updating of information on discharges and their causes; (b) closer relations between environmental authorities and regulated entities; and (c) growing awareness of natural resources among users, with reduction of industrial discharges being a concrete result of implementation.

The above elements were discussed in the process of modifying Decree 901 that led to Decree 3100 being issued with amendments to the design of the water pollution charge. Once the transitory period for applying the modifications is completed, it will be possible to determine if the problems were solved.

15.5.2 Recommendations for Developing Countries

The Colombian water pollution charge programme constitutes a valuable case study for analysis of the various elements for implementation of an environmental economic instrument in developing countries, especially the application of environmental charge to control pollution.

The following elements are prerequisite that allow the development of the charge:

1. The existence of a strong and well-designed environmental institutional framework in which the institutions have a high profile at central government and regional levels. Management of the environment at ministerial level with specialised environmental authorities in the regions provides a mix of independence and specialisation that enables enforcement a pollution charge.
2. The constitutional and legal basis for the pollution charge are important and essential conditions. The constitution establishes the differences between a tax and a charge. The tax will be set by the local and national administrations, generally to finance all public programmes and institutions. On the other hand, a charge is levied because of a service that the government is delivering to the user. An environmental charge could be set because the government is responsible for administering environmental resources. Specifically in the case of pollution, the water is being used for waste disposal, and the government must guarantee the sustainability of the water resources. The charge is legally based to recover the cost to the government of managing the water resources. The revenue has to be directed to the service itself, and does not go to the general budget of the country. This distinction allow for design of the pollution charge as an economic instead of a purely finance instrument.
3. The Minister of the Environment and directors of the regional authorities have the political will to implement an unpopular measure.
4. The continuation of the policy at both technical and political levels is essential.

Specifically following recommendations are made for the design and implementation of a pollution charge based on the Colombian experience:

1. In general terms, the Baumol and Oates (1988) approach is considered practical and efficient in setting a pollution charge. The setting of the environmental goal is a key element, not only for the regulator but also for the acceptability of the charge by the sources and the community.
2. The periodic evaluation of the environmental goal and making public such evaluations is also an important issue. The goal must also include monitoring of the quality of the water resources as well as the discharges.
3. The period of evaluation of the goal and the tariff adjustments have to be carefully designed in order to allow the sources of pollution to take the proper decisions and implement their abatement measures.
4. The level of the tariff is not a critical aspect as long as it is not set so high as to threaten sources with bankruptcy, or force them not to pay. It is better to start with a relatively low tariff and adjust gradually to meet the goals.

5. Adjustment of the tariff, in the Colombian case, provides an effective incentive for promotion of discharge reductions.
6. The implementation of the charge has to take into consideration the industries or activities that are regulated by the government and their interactions with the charge.
7. Although the revenues of the charge can help to finance the environmental institutions, it is politically critical to have transparent mechanisms for the investment of resources and to allow the sources and the community to supervise the effectiveness and impacts of the investments.

15.6 References

Baumol W, Oates W (1971) The Use of Standards and Prices for Environmental Protection. The Swedish Journal of Economics 73:42-54

Baumol W, Oates W (1988) The Theory of Environmental Policy. 2nd edn, Cambridge Press, Cambridge, UK

Castro LF, Castro R (2002) Tasas Retributivas por Vertimientos Puntuales. Evaluación Nacional. Ministerio del Medio Ambiente, Bogotá, Colombia

Castro LF, Caicedo JC, Jaramillo A, Morera L (2002) Aplicación del principio contaminador-pagador en América Latina: evaluación de la efectividad ambiental y eficiencia económica de la tasa por contaminación hídrica en el sector industrial colombiano. Serie Medio Ambiente y Desarrollo – Economic Comisión for Latin America and the Caribean – ECLAC, United Nations, Santiago de Chile, Chile

Contraloría General de la República (2003) Estado de los Recursos Naturales y del Ambiente 2002 – 2003. Imprenta Nacional, Bogotá, Colombia

Corporación Autónoma Regional de Cundinamarca - CAR (1993) Estudio para la Estrategia de Saneamiento del Río Bogotá. Bogotá, Colombia

Departamento Nacional de Planeación, Ministerio del Medio Ambiente, Ministerio de Desarrollo (2002) Acciones Prioritarias y Lineamientos para la Formulación del Plan Nacional de Manejo de Aguas Residuales. Documentos Consejo Nacional de Política Económica y Social - CONPES 3177, Bogotá, Colombia, 2002

Field B (1995) Economía Ambiental: Una Introducción. Mc Graw Hill, Colombia

Ministerio del Medio Ambiente (1998a) Fondos Regionales de Inversión en Descontaminación Hídrica, Bogotá, Colombia

Ministerio del Medio Ambiente (1998b) El que Contamina Paga: Aguas Limpias para Colombia al Menor Costo. Implementación de las Tasas Retributivas por Contaminación Hídrica. Imprenta Nacional, Bogotá, Colombia

Ministerio del Medio Ambiente, CARDER, CVC, CORNARE (1998) Programa de Cooperación Horizontal de Tasas Retributivas por Vertimientos Puntuales. Informe Final Fase I. Bogotá, Colombia

Ministry of the Environment (2003). http://www.minambiente.gov.co

OECD (1994) Managing the Environment: The Role of Economic Instruments. Organisation for Economic Co-operation and Development, Paris

World Bank (2000) Greening Industry. New Roles for Communities, Markets and Governments. Oxford University Press, USA

World Bank and Ministry of the Environment (1997) Colombia's Pollution Charge Program: An evaluation with recommendations for implementation." Seminar on Evaluation and Implementation of the Pollution Charge Program, Bogotá, Colombia, April 28

Index

Printing: Krips bv, Meppel
Binding: Stürtz, Würzburg